Routledge Research in Character and Virtue Education

This series provides a forum for established and emerging scholars to discuss the latest debates, research and theory relating to virtue education, character education and value education.

The Theory and Practice of Virtue Education
Tom Harrison and David Walker

Cultivating Moral Character and Virtue in Professional Practice
David Carr

Virtues in the Public Sphere
Citizenship, Civic Friendship and Duty
James Arthur

Virtues and Virtue Education in Theory and Practice
Are Virtues Local or Universal?
Catherine A. Darnell and Kristján Kristjánsson

Virtues and Virtue Education in Theory and Practice

Are Virtues Local or Universal?

Edited by Catherine A. Darnell and Kristján Kristjánsson

Routledge
Taylor & Francis Group
LONDON AND NEW YORK

First published 2021
by Routledge
2 Park Square, Milton Park, Abingdon, Oxon OX14 4RN

and by Routledge
52 Vanderbilt Avenue, New York, NY 10017

Routledge is an imprint of the Taylor & Francis Group, an informa business

British Library Cataloguing-in-Publication Data
A catalogue record for this book is available from the British Library

Library of Congress Cataloging-in-Publication Data
Names: Darnell, Catherine A., editor. | Kristjánsson, Kristján, 1959– editor.
Title: Virtues and virtue education in theory and practice : are virtues local or universal? / Edited by Catherine A. Darnell and Kristján Kristjánsson.
Description: Abingdon, Oxon ; New York, NY : Routledge, 2021. | Series: Routledge research in character and virtue education | Includes bibliographical references and index.
Identifiers: LCCN 2020024216 (print) | LCCN 2020024217 (ebook) | ISBN 9780367356491 (hardback) | ISBN 9780429343131 (ebook)
Subjects: LCSH: Virtues—Study and teaching. | Moral education. | Education—Philosophy. | Educational psychology.
Classification: LCC BJ1531 .V59 2021 (print) | LCC BJ1531 (ebook) | DDC 179/.9071—dc23
LC record available at https://lccn.loc.gov/2020024216
LC ebook record available at https://lccn.loc.gov/2020024217

ISBN: 978-0-367-35649-1 (hbk)
ISBN: 978-0-429-34313-1 (ebk)

Typeset in Sabon
by Apex CoVantage, LLC

For Tom and Dexter (CAD)

Contents

Tables

Acknowledgements

This edited volume has its origin in a conference entitled "Virtues and Virtue Education: Local or Universal", organised by the Jubilee Centre for Character and Virtues, where both the present editors worked at the time, and held at Oriel College Oxford in January 2019. The current line-up of papers was chosen from a range of over 70 presentations; we are happy to announce that almost all the authors we contacted agreed to contribute.

We are grateful to the main organiser of the conference, Professor James Arthur, for the suggestion that we turn the best papers into a volume. We are equally thankful to our Routledge editor, Emilie Coin, for her support of our project from the very outset, and to her two academic reviewers for helpful and incisive comments on our book proposal. Most of all, however, we are grateful to the contributors for respecting relevant deadlines, responding enthusiastically and promptly to our email correspondences, and taking well to our critical comments on their first drafts. It was a pleasure to work with them all.

We must also give a big thank you to Matthew Collins, a Research Assistant at the Jubilee Centre for Character and Virtues, who helped us copy-edit the final version of the manuscript.

This volume in no way presents an exhaustive coverage of topics or viewpoints related to its broad theme. However, we hope that it provides a source of insights and inspirations, not least for young researchers entering the field of virtue and character research – be it from the entry point of psychology, philosophy, or education – and that it helps move the discourse at least a few steps forward.

Birmingham, April 2020,
Catherine A. Darnell
Kristján Kristjánsson

Biographies

Editors

Catherine A. Darnell, PhD, is a Lecturer in Cognitive Psychology in the School of Education at The University of Birmingham. Her interests lie in cognitive development particularly metacognition, decision making, and the impact of these on learning. Her most recent work has explored moral decision making in adolescents and adults.

Kristján Kristjánsson (PhD, University of St. Andrews) is Deputy Director in the Jubilee Centre for Character and Virtues and Professor of Character Education and Virtue Ethics, University of Birmingham, U.K. His interests lie in research on character and virtues at the intersection between moral philosophy, moral psychology, and moral education.

Contributors

Sarah Banks is Professor in the Department of Sociology and Co-director of the Centre for Social Justice and Community Action, Durham University, UK. She teaches and researches in the fields of social, community and youth work, and has published texts on professional ethics, including *Ethics and Values in Social Work* (Palgrave Macmillan, 2012) and *Ethics in Professional Life: Virtues for Health and Social Care* (with Ann Gallagher, Palgrave Macmillan, 2009).

Ron Beadle is Professor of Organisation and Business Ethics at Northumbria University, UK. His research aims to defend, extend and apply Alasdair MacIntyre's Thomistic Aristotelianism in organisations. He has published in *Business Ethics Quarterly*, *American Catholic Philosophical Quarterly*, and *Organisation Studies*, amongst others. Ron's empirical research focuses on travelling circuses.

Yi-Lin Chen, PhD, gained a doctorate from the Institute of Education in London in 2010. Yi-Lin is currently an Associate Professor in the Department of Education at National Taiwan Normal University, Taiwan. Yi-Lin's research interests are in virtue ethics and character education, social and emotional learning, and teacher education.

Randall Curren is Professor and Chair of Philosophy and Professor of Education (secondary) at the University of Rochester, and an Honorary Senior Research Fellow at the Jubilee Centre for Character and Virtues at the University of Birmingham. His work spans Ancient Greek philosophy, ethics, and philosophy of education.

Doret de Ruyter is Professor of Philosophy of Education at the University of Humanistic Studies since June 2018. Previously, she held the same position at VU Amsterdam for 13 years. Her research interests centre on flourishing as an aim of education, moral education, and citizenship education. She is currently president of the Association for Moral Education.

William Fleeson is a Professor of Psychology at Wake Forest University. He received a BA in philosophy from Wisconsin and completed a PhD in psychology from Michigan. He completed postdoctoral training in development from Germany. William's interests include moral exceptionality, the nature of personality, borderline personality disorder, consistency, and self-regulation. He takes a personality, social, developmental, and philosophical approach.

Jonathan Jacobs is Professor of Philosophy, and Director of the Institute for Criminal Justice Ethics at John Jay College/CUNY. He is also on the doctoral faculties of Philosophy and Criminal Justice at the CUNY Graduate Center. His interests include moral psychology, moral epistemology, and relations between ethics, politics, and criminal justice.

Eranda Jayawickreme, PhD, is Associate Professor of Psychology and Senior Research Fellow at the Program for Leadership and Character at Wake Forest University. His research focuses on post-traumatic growth, moral personality, wisdom, and well-being. His awards include the 2018 Faculty Excellence in Research Award from Wake Forest, the 2015 Rising Star award from the Association for Psychological Science, a Mellon Refugee Initiative Fund Fellowship, and multiple grants from the John Templeton Foundation, Templeton Religion Trust, Templeton World Charity Foundation, European Association for Personality Psychology, and the Asia Foundation/USAID.

James MacAllister, PhD, is a Senior Lecturer in Philosophy of Education, The University of Edinburgh. His publications include a monograph called *Reclaiming Discipline for Education*, as well as various articles on Scottish philosophers and education. He is currently working on a second monograph called *Educational Institutions and the Common Good.*

Christian B. Miller is the A. C. Reid Professor of Philosophy and Past Director of the Character Project at Wake Forest University. He mainly works on virtue and character and has just completed a book manuscript on the virtue of honesty.

Elise D. Murray, PhD, recently received her doctorate from Tufts University and is a Senior Research Scholar in the Program for Leadership and Character at Wake Forest University. She assesses character programming to promote thriving in young adults and has a particular interest in the development of intellectual humility.

Mike Prentice, PhD, received a PhD in Social and Personality Psychology from the University of Missouri, where he received the Distinguished Dissertation in the Social Sciences. He was recently a Postdoctoral Research Associate of the Beacon Project at Wake Forest University.

Angus Robson, PhD is a Senior Lecturer in Business Ethics and Leadership at Northumbria University, UK. His main research interest is in Aristotelian virtue ethics, with an emphasis on hermeneutics and the work of Alasdair MacIntyre. He has published on ethics in the contexts of banking and care work.

Jennifer Rothschild, PhD, is Assistant Professor of Philosophy at the University of Florida. Her primary area of research is Aristotelian virtue ethics.

Nancy E. Snow, PhD is a Professor of Philosophy and Director, Institute for the Study of Human Flourishing. She is the co-author (with Jennifer Cole Wright and Michael Warren) of a book on virtue measurement (currently under review); is the series editor of a multidisciplinary 15-volume series entitled "The Virtues", for Oxford University Press; is currently at work on a number of invited papers; is working on a book on hope as a democratic civic virtue; and plans to write a book on virtue ethics and virtue epistemology.

Jonathan M. Tirrell, PhD, is a Research Assistant Professor in the Institute for Applied Research in Youth Development at Tufts University. He currently works on the Compassion International Study of Positive Youth Development, a longitudinal evaluation of Compassion International's youth-development programs aimed at promoting thriving among youth living in poverty.

Jonathan Webber is Professor of Philosophy at Cardiff University. He has published on moral psychology and virtue ethics in leading journals, including *Analysis*, *European Journal of Philosophy*, *Journal of Moral Philosophy*, *Mind*, and *Philosophical Quarterly*. He focuses on how findings in the social psychology of attitudes should inform virtue ethics.

Dieter Weber is Professor for Ethics in Social Work at the Faculty of Diaconic Studies, Health Care and Social Work at the University of Applied Sciences and Arts Hanover, Germany. Research interests lie in theory of recognition, theory of gift, phronesis, and reflective judgement as a basis for professional knowledge for social work.

Jennifer Cole Wright, PhD, is an Associate Professor in Psychology at the College of Charleston. She studies virtue, meta-ethics, moral conviction, and tolerance, the influence of individual and social mindsets on moral judgments, and young children's early moral development. Her most recent publications include *Understanding Virtue: Theory and Measurement* (with Michael Warren and Nancy Snow) and has edited an interdisciplinary volume on *Humility* (both with Oxford Press), as well as co-editing, with Hagop Sarkissian, *Advances in Experimental Moral Psychology.*

Heidrun Wulfekühler is Professor for Ethics in Social Work at the Faculty of Diaconic Studies, Health Care and Social Work at the University of Applied Sciences and Arts Hanover, Germany. Her main interests in teaching and research are virtue ethics in social work and the didactics of ethics.

Introduction to the volume

The central aim of this volume is to bring into sharper focus debates about the universality versus the locality of virtues, within both philosophy and psychology, and the relevance of those debates for the practice of virtue education, or "character education", as it is more commonly called. Since most of Western virtue theory harks back to Aristotle, it is *de rigueur* to begin a review of those debates with a nod back to him. However, although Aristotle is typically taken to come down on the universality side of the debates, his position is neither as simple as it may appear at first sight nor amenable to a single interpretation.

To be sure, as Stangl correctly points out (2018, p. 511), Aristotle aspired to give a single objective account of flourishing, and for him flourishing is primarily constituted by the continuous and stable activities inherent in moral, civic, and intellectual virtues – which would normally be seen as objective also and presumably then as somehow universal, given that we are talking about one unique natural species: humankind.[1] *Gens una sumus*: we are one people. However, Aristotle complicated this picture with his view of the radical individuality-adjustment of virtues to *developmental level*, *social position*, and *individual constitution*. For example, emulousness is a virtue for young people whereas adults do not need to emulate role models. Magnificence and magnanimity are virtues for people blessed with unusually abundant material resources but not for ordinary folks. Temperance in eating is not the same for Milo the athlete as for the university professor, because what is intermediate in virtue is relative to the individual, "not in the object" (Aristotle, 1985, p. 1106b1–7). And, from an educational perspective, a boxing instructor will not "impose the same way of fighting on everyone" (1985, p. 1180b9–11). There is thus no *one* best way across individuals to be, say, virtuously generous as opposed to be being stingy or wasteful. It all depends on your individual circumstance (are you poor or wealthy?) and your own natural inclination towards either extreme, away from which you should try to drag yourself, with the help of friends. Hence, "it is clear that the good cannot be some common [nature of good things] that is universal and single" (1985, p. 1096a25–9), and "each state [of character] has its own special [view of] what is fine and pleasant" (1985, p. 1113b31–3). This does not apply only to representations of individual virtues. How different virtues cluster and gain greater expression and dominance in each individual's life also depends on the individuality of the virtuous person. There is thus no general master moral virtue in Aristotle's theory, although there is an intellectual meta-virtue for integration and adjudication: *phronesis* (Darnell, Gulliford, Kristjánsson, & Panos, 2019).

To put this in simple terms, Aristotle's individuality-adjustment thesis proposes that there is no description of an individual's virtue repertoire available to us that abstracts from its instantiation in that person in all her psychosocial uniqueness. There is no single unique blueprint of the perfectly virtuous person, per se, to aim for. When I love the generosity of my friend, I do not love it as matching well or less well the repeatable generosity of the perfectly generous person, for there is no such generosity *simpliciter*. There is an endless plurality of traits that all make the grade as virtuous generosity, instantiated in different persons, and there is no way to choose between them in ways that satisfies some ideal condition of virtue maximisation (see further in Kristjánsson, 2020).

Let us, for the sake of argument, grant these points. Prior to the issue of individuality-adjustment, however, there lurks a deeper philosophical question about the nature of the virtues that, subsequently, need to be adjusted to individual and societal circumstances. Are *those virtues* local or universal? It is this question, precisely, that occupies the minds of most of the authors in Section 1 of our volume, and indeed some authors in the two other sections, too. The problem is that Aristotle is not entirely unambiguous on this prior question. The standard naturalistic reading of him by philosophers such as Nussbaum, Foot, and Hursthouse – which sees him as evaluating the goodness of virtues through the same structure as evaluating the goodness of features of plants and animals – highlights the universal nature of human virtues. This is the sort of Aristotle who says, in a much-repeated quote, that the more we travel, "we can see how every human being is akin . . . to a human being" (1985, p. 1155a20–2). Yet in other places, especially in his *Politics*, Aristotle admits cavalierly that virtues such as justice are relative to political constitutions in different city states, and he does not seem to be expressing this as a descriptive anthropological statement only.

Famously, the person who is perhaps most responsible for the resurgence of interest in Aristotelian virtue theory, at least outside the charmed circles of academic philosophy, Alasdair MacIntyre (1981), offered an interpretation of Aristotle that is much more amenable to a localised, if not a relativised, interpretation: namely an account of Aristotelian virtue that fits within a broader historically and socially situated theory of goods as internal to particular practices, traditions, and cultures. However, MacIntyre is not always easy to pin down; even in his 1981 book, he did not say outright that the rationale of virtues was fully localised; and in subsequent works he has moved more into the direction of arguing that different virtue traditions can be qualitatively compared according to objective criteria, culminating in his latest book (MacIntyre, 2016, pp. 222–223) where he waxes almost as naturalistic as Nussbaum about certain categories of goods (and the virtues attached to them) contributing to the good life of human beings wherever you go and at whatever time in history.

In his original contribution (1981) to this debate, MacIntyre had singled out a certain kind of "heroic society" (found especially in Ancient Greece and Viking-time Iceland, at least as depicted in the Sagas) that presented unique characteristics and discrete virtues, not likely to feature as such in any other type of culture. This dichotomy mirrors another common one, which in some cases draws on MacIntyrean insights but in others predates his work, between two divergent sorts of human societies with radically different sorts of virtue repertoires: individualist and collectivist (sometimes referred to as "Western" versus "Eastern", although that terminology is overly simplistic). Markus and Kitayama's (1991) construal of *interdependent* (Eastern/traditional/collectivist) versus *independent* (Western/liberal/individualist) self-conceptions thus undoubtedly constitutes one of the most influential works in late twentieth-century social psychology, with far-reaching implications in many areas, including moral philosophy, emotion theory, and character education. These conceptions, it turns out, involve not only different psychological self-images, but also radically divergent ways of feeling, seeing, acting, and being in the world – and they give rise to contrasting sets of virtues. Markus and Kitayama's juxtapositioning of the Western-bounded, unique, self-contained, and segregated liberal self-concept with the Eastern fluid, connected, holistic, and less-differentiated traditional self-concept is based on a whole mountain of social scientific research. Since the publication of their article, mountains more have been added of analyses and supporting (and, in some cases, conflicting) findings (Kristjánsson, 2010, Chapter 8). Unfortunately, this particular debate does not feature at all in the present volume, as noted by one of our reviewers. That is understandable, however, because it has gradually become isolated within certain discursive silos in social psychology and sociology, much to the detriment of more general academic debates about the possible relativity of virtues or other moral constructs. Despite the interdisciplinarity of the present volume, and the host of different voices it represents, most of those (especially in Section 1) address the question of universality versus locality of virtues in terms that are familiar from standard philosophical debates about the issue.

Speaking of those standard philosophical debates, Rebecca Stangl (2018) has, so to speak, stolen our thunder with her excellent and almost exhaustive overview of the nooks and crannies, pros and cons, of the different philosophical approaches in question. Rather than replicating her argumentation, we will, in each of the three introductory chapters to the three sections of the book, focus on the issues that are most immediately relevant to the debates that play out in those sections.

Note

1 We are talking here about general "human" virtues. Everyone will agree that professional virtues differ, sometimes to a substantial degree, between professions.

References

Aristotle (1985). *Nicomachean ethics* (T. Irwin, Trans.). Indianapolis: Hackett Publishing.

Darnell, C., Gulliford, L., Kristjánsson, K., & Panos, P. (2019). *Phronesis* and the knowledge – action gap in moral psychology and moral education: A new synthesis? *Human Development*, *62*(3), 101–129.

Kristjánsson, K. (2010). *The self and its emotions*. Cambridge: Cambridge University Press.

Kristjánsson, K. (2020). Grounding deep friendships: Reconciling the moralized and aestheticized views. *Journal of Philosophical Research*, in press.

MacIntyre, A. (1981). *After virtue*. Notre Dame: University of Notre Dame Press.

MacIntyre, A. (2016). *Ethics in the conflicts of modernity*. Cambridge: Cambridge University Press.

Markus, H. R., & Kitayama, S. (1991). Culture and self: Implications for cognition, emotion, and motivation. *Psychological Review*, *98*(2), 224–253.

Stangl, R. (2018). Cultural relativity and justification. In N. E. Snow (Ed.), *The Oxford handbook of virtue* (pp. 508–523). Oxford: Oxford University Press.

Section 1

Philosophical issues

One of the reviewers of our book proposal complained that most of the contributors to the volume endorsed some variety of the universality thesis. While few of the contributors are as explicit as Jacobs (Chapter 2 of this volume) in stating that "virtue-centred ethics should be understood non-relativistically, non-locally" and even as providing "reasons to reject relativism", it is true that the sort of academics who turn up at conferences on virtue ethics, or its educational incarnation as character education, are rarely followers of moral relativism in any form or guise. Nevertheless, most of the present contributors make even-handed attempts at justifying their positions with respect to opposing, in particular relativistic, views – so the latter do get a fair hearing in many chapters. Moreover, the whole idea behind the motivation to justify some kind of a universality thesis, radical or modified, cannot be fully understood except against the backdrop of the observation that the norms endorsed by various cultures undeniably differ to a substantial degree (cf. the opening sentence in Stangl's overview article, 2018, p. 508). Thus, in a way, the universalism thesis already contains an attempt to challenge what could be called "the received wisdom"; and received wisdoms rarely needs as much spelling out as the arguments meant to counter them.

That said, it may again be an over-simplification to say that a thesis about the locality of virtues constitutes an undeniable received wisdom (the commonly encountered, developmentally induced relativism of high-school students and undergrads notwithstanding!). Offsetting any insights about cultural uniqueness – be those modest insights about how the British mentality differs subtly from that of its mainland European counterparts, or more radical ones about the extreme opaqueness and unfamiliarity of cultures such as that of the Amish people – are more mundane observations about how surprisingly similar human beings seem to be in some respects. Indeed, it would probably seem astounding to an alien from outer space how monocultural human beings are, irrespective of distances in geography and time. We more or less like and do the same things: eating, making love, playing games, creating art, socialising, gossiping, etc. We even started to build pyramids on both sides of the Atlantic at about the same time in history without any cross-Atlantic interaction. It would be somewhat astounding if our repertoire of virtues did not reflect those similarities in any way.

Recent historical and political developments have brought some of those considerations into sharper relief and, indeed, provided some of the motivations for the theme of the Oxford conference from which this volume derives. We are here obviously referring to the growing political polarisation in the Western world and the recent upsurge of post-truth populism and nationalism. We seem to have moved far away from the cosy consensus of the 1990s when a liberalist mindset, in one form or another, seemed to have become the norm in Western socio-political debate and writers mused profusely on the "end of history", in the sense that the time of radical ideological tensions was gradually drawing to a close (in the Western world and soon elsewhere, too), with the victory of soft capitalist economic structures and a social liberalist consensus. What nobody anticipated, but has happened since then, is what the columnist O'Hagan (2020) refers to as an "anti-woke backlash", with a silent majority of an odd mixture of privileged and under-privileged voters reacting badly to what they see as excesses of censorious political correctness, watered-down local traditions, bland and homogeneous "luvvie" culture, and anything-goes social mores.

To see recent political developments as an affirmation of local over global values and virtues is true to some extent. However, the alleged "liberal consensus" was not exactly built around the fulcrum of universalist virtues either, except perhaps one: *tolerance*. However, tolerance is a somewhat paradoxical virtue from the perspective of the debate about locality versus universality, because tolerance affirms the universal acceptance of difference. After all, liberalism rejects any thick comprehensive conception of the good life, so liberalist philosophy is not exactly the ideal home for a conception of virtues at all, let alone universalist ones. And although the current socio-political turmoil served as the backdrop of many of the presentations at the 2019 Oriel conference, the discussion quickly turned to more specific and time-honoured philosophical conundrums.

In Chapter 1 of this volume, Miller offers a helpful discussion of different kinds of relativism and notes that the one which need concern us in the present context is the *meta-ethical* version. He further explains the three levels at which meta-ethical misgivings about the universality of virtue may emerge. The *first* and most general one is about the presumed universal domains of life that the discrete virtues are meant to track. What exactly are those domains and their related virtues – and why is there no intercultural agreement about the exact lists of either two? *Second*, going down the ladder, even if we agree on a particular domain as having universal salience, say the domain of self-worth, what is the virtue relevant to that domain: is it pride or is it humility? These are conflicting virtues, upheld by different theorists and cultures. *Third*, even if we agree on a domain and its respective virtue – say, the domain of responding to facts being governed by the virtue of honesty – there tend to be conflicting accounts of what exactly the virtue involves, within cultures and even more so between cultures, as soon as we attempt to populate the relevant "thin" virtue term with sufficient "thickness" to constitute an action-and-emotion guiding state of character. Miller could even have added a *fourth* level: what happens when this specific virtue clashes with another one and *phronesis* is meant to kick in for adjudication? Does the context-sensitive nature of the *phronetic* decision, ideally to be arrived at in each case, mean that *phronesis* cannot operate along universalist principles? Moreover, if we agree that different decisions can be equally appropriate after the relevant individuality-and-context adjustments, is such pluralism compatible with universalism, or does it push us in the direction of relativistic localism?

In Chapter 4, Webber provides us with a helpful conceptualisation that complements Miller's insights. After explaining how the motivational structure of each virtue consists in a cluster of related evaluative attitudes, Webber employs the metaphor of a high and low resolutions. Our ordinary virtue talk "specifies complex motivational states in fairly low resolution", allowing for a universalist interpretation. However, if we zoom in for a higher-resolution image that presents the relevant set of attitudes, or even further still for an even higher-resolution image that presents the evaluative attitudes' clusters of associated desires and beliefs, we end up with much more localised (culture-specific or even situation-specific) variance. However, Webber (as indeed Miller also, while using different descriptors) allows for the possibility that this "zooming" exercise may yield different results in the case of different virtues: namely, that some virtues are universal "all the way down" while others lose their universalist "resolution" fairly quickly as we zoom in on the specifics. Both Miller and Webber offer suggestions that would potentially change the constitution of the discourse animating the present volume from being about the universality or locality of *virtue* to being about the universality or locality of *particular virtues*.

In addition to this general debate, to which all the chapters in Section 1 contribute in one way or another, each chapter offers its unique take on particular issues. Miller's case study of the surprisingly neglected virtue of *honesty* indicates that its universality remains even in Webber's high-resolution mode. Webber makes a similar claim for *ethical integrity*, because the attitudes composing it do not refer to the particular ethical environment at all. In Chapter 2, Jacobs proposes a universalist argument from the nature of good moral education. This is quite a novel take on the universalism issue because claims about the universality of good moral education tend to be derived from the standpoint of some general ethical theory which the relevant theorist endorses and, subsequently, argues should inform moral education. Jacobs turns the argument on its head. He argues that moral education which cultivates the virtues (normally referred to as character education) is the sort of education that is "best able to facilitate comprehension of moral values, their complex interrelations, and responsiveness to them". Indeed, this is the way most parents and educators around the world do teach children in order to help them respond to the texture of moral reality. If that is the case, however, we have good reason to think that the moral reality to which the children are being alerted is indeed virtue-based. Moreover, since the methods to teach character tend to be fairly uniform, this gives us reason to think that the virtues that are being cultivated are also universal.

In Chapter 5, Rothschild argues that the "core" Aristotelian moral virtues such as justice, generosity, moderation and courage (although she excludes magnanimity), as well as some that Aristotle did not name, such as honesty and patience, are fundamentally social. This essential sociality of the core virtues, with their common existential tasks, entails that they make us good in ways that render them "fully universal": namely, "relevant to all of us, worth pursuing for each person in pursuit of the good, and in particular, for the good life enjoyed jointly with others". Rothschild acknowledges, however, the existence of other more local "secondary virtues". In contrast to the upbeat tone of Rothschild's chapter, in Chapter 6, Snow brings us down to earth with her admonitions to virtue theorists for not staying in touch with recent developments in evolutionary science in general and in germline genome editing in particular: a technology that can produce heritable changes in members of many species, including humans. Virtue theorists tend to rely on low-level anecdotal evidence from everyday life but shun real science. The traits that can potentially be altered are not only the standard biological ones, having to do with disease prevention and an extended life span, but may also include moral traits and capacities. What would be the moral implications of this technology becoming widely available (or perhaps only available to a select few), and how would it problematise the whole universality–locality discourse? In Chapter 3, Curren also takes virtue theorists to task in a way that relates to the concerns of this volume. His worries are not about the evasion of science, but rather about the evasion of talk about universal rules and laws, as enshrined for example in the natural-law tradition. Rather than drawing on Aristotle's *Politics*, as he has done in previous work, Curren here challenges a standard interpretation of Socrates as a virtue ethicist with no interest in moral law and duty. Rather, he depicts a philosopher whose understanding of virtue is "informed by natural law and a related understanding of piety and *fidelity to reason*". Curren's discussion elicits various vexing questions about the role of laws and rules in a virtue-based system, and whether a debate on the universality of the latter can be conducted without an account of the universality of the former.

So what are the lessons to be learned from Section 1 of this volume, apart from evidencing how rich and illuminating – yet varied – the accounts are that these authors give us? There are many lessons, but we leave readers with just one. Perhaps the discourse on the universality of virtue has been conducted so far at too high a level of abstraction. We might be better served with more detailed accounts of individual virtues and how those represent universality or locality – but without assuming that one general answer can found which applies to the virtue repertoire as a whole.

References

O'Hagan, E. M. (2020). The "anti-woke" backlash is no joke – and progressives are going to lose if they don't wise up. *The Guardian*. Retrieved January 30, 2020, from www.theguardian.com/commentisfree/2020/jan/30/anti-woke-backlash-liberalism-laurence-fox

Stangl, R. (2018). Cultural relativity and justification. In N. E. Snow (Ed.), *The Oxford handbook of virtue* (pp. 508–523). Oxford: Oxford University Press.

1 Moral relativism and virtue

Christian B. Miller

Introduction

While there is an extensive literature on moral relativism in meta-ethics, little has been said in assessing the view with respect to virtue specifically. The goal of this chapter is to provide an overview of what I take to be the central issues surrounding moral relativism and moral virtue. Hence, rather than trying to advance a moral relativist or moral realist position in detail myself, my approach is more programmatic. I also try to avoid too much technicality in the hope that the chapter will be of interest and relevance to those outside the narrow halls of contemporary analytic philosophy.

Here is my plan. In the first section of this chapter, I review the three main forms of moral relativism – descriptive, normative, and meta-ethical – before focusing just on the third. From there, the central issue becomes what concerns arise which pertain to relativism and virtue specifically. In section two, I set aside three potential concerns which do *not* arise, having to do with rejecting abstract rules, accepting the context sensitivity of virtue, and adopting moral particularism. In section three, we will see that there are three other legitimate concerns instead having to do with the correct list of virtues, the choice of particular virtues, and the account of a given virtue. I assess the severity of each of these concerns. The chapter ends with a case study of the virtue of honesty. I provide an initial account of the virtue and suggest that a relativist approach does not seem promising at least for this virtue.

Background on moral relativism

There is no one view which goes by the name of "moral relativism". In fact, the label applies to several different kinds of moral relativism (hereafter MR), none of which entails the other. The three leading kinds are descriptive, normative, and meta-ethical moral relativism. The focus of the chapter will be on the last kind, but let me briefly introduce the other two.[1]

Descriptive moral relativism is a view about the extent of differences in what people believe about moral standards. It could pertain to both differences between individuals and to differences between societies and cultures. The following, for instance, is a very weak version of the view:

Weak descriptive MR

There currently exist some moral agents A and B, each of whom endorses a set of moral standards such that:

(1) The moral standards in each set are not logically equivalent.
(2) A and B would disagree with each other in some of their moral evaluations as a result of applying their divergent sets of moral standards.

Note that this is an anthropological position, not a traditional philosophical position. There is no way to assess the plausibility of versions of descriptive MR *a priori* from the armchair.

Clearly, at least this weak version of the view is true. Once the view is strengthened by, for instance, increasing the number of individuals with opposing moral standards, or increasing the number of standards for which there is disagreement (or both), then it becomes more controversial (for additional discussion, see Miller, 2002).

Normative MR is significantly different. It is roughly the view that individuals should be highly tolerant towards the behaviour of people with different moral frameworks. Here, for instance, is one way of developing the view:[2]

Strong normative MR

For any two individuals A and B who hold substantially different sets of moral standards,

(1) A should never judge the moral behavior of B in light of A's moral standards, unless they happen to overlap with B's moral standards.
(2) A should never act toward B in such a way as to try to bring B into conformity with A's moral standards.
(3) Similarly, for B's judgments and behavior toward A.

In this case, we do not have an anthropological view but rather a straightforwardly moral position. In other words, the normative moral relativist is making moral claims about what people should and should not think and do.

Clearly there will be some weak versions of normative MR that will be widely accepted. But for the most part, substantive versions of the view are highly controversial and often rejected (for additional discussion, see Miller, 2002).

Finally, we come to meta-ethical moral relativism. One central commitment is a negative claim.

No Objectivity Thesis

> There are no objective moral truths.

This thesis is intended to reject a moral realist approach, according to which moral truths both exist and are mind-independent. What this amounts to is roughly the view that the truth of the moral propositions in question does not depend upon the views and attitudes towards those truths held by human beings. They would be true, even if human beings thought otherwise or did not have any views about them in the first place.[3]

But merely rejecting the existence of objective moral truths does not make one a meta-ethical moral relativist. Rival positions such as traditional emotivism and the moral error theory also accept the No Objectivity Thesis. So in addition, let us add the following.

Framework Relativity Thesis

> For any individual, if that individual accepts some moral claim, then the truth of that claim is fixed only by that individual's actual moral standards and any descriptive facts relevant to the claim.

For instance, suppose that Jones judges that Smith deserves the death penalty. If Jones believes that the death penalty is morally obligatory for crimes like serial murder, and if Jones also believes that Smith committed serial murder, then that is sufficient for its being true (for Jones) that Smith should be executed. Tahani might have a different moral code such that the death penalty is always morally wrong. Then when she judges that Jones, despite being a serial murderer, should not be executed, that judgment turns out to be true as well (for Tahani).

Note that meta-ethical MR is understood here in terms of the individual's *actual* moral standards. There are versions of constructivist positions in meta-ethics which focus on the individual's *improved* moral standards, but which are still relativist views. Fortunately for my purposes in this chapter, we do not need a formulation of meta-ethical MR which casts its net perfectly. The previous will do the work we need (see also, Miller, 2009).

One final observation is that, as its name suggests, meta-ethical MR is not an ethical position. Rather, it is a metaphysical or ontological position, i.e., a view from outside of ethics about the metaphysical status of ethical truths. One could endorse this view and also accept a Kantian approach, a utilitarian approach, a virtue ethical approach, or any other approach in first-order normative ethics. By itself, it is neutral on substantive ethical debates at the level of normative ethics and also applied ethics.

How plausible is meta-ethical MR? As with every position, it can be formulated in different ways that range in their plausibility. But it is fair to say that standard formulations are widely rejected within contemporary analytic philosophy. Rival moral realist and expressivist views, for instance, have a much larger following. But popularity does not determine truth, and as far as this chapter is concerned, I will remain neutral on the overall plausibility of meta-ethical MR (see also, Miller, 2002, 2009, 2011)

In what follows, this is the form of moral relativism that will primarily be discussed in relation to virtue and character. When philosophers discuss relativism and virtue, it is the meta-ethical version that they have in mind. Normative MR will be set aside completely here and would need to be discussed separately in relation to the virtue of tolerance. Descriptive MR will be mentioned at various points only in passing, since it is an anthropological approach and not in the first instance a philosophical approach. Hence, subsequent reference to "moral relativism" should be understood to refer just to the meta-ethical version of the view.

Three concerns which do not arise

Before exploring the relationship between meta-ethical MR and virtue, a few preliminary points are worth making. First, my interest here is in *virtue theory*, rather than *virtue ethics*. According to this now familiar distinction (e.g., Snow, 2010), virtue theory is concerned with giving an account of the nature of virtue and of particular virtues. Virtue ethics, in contrast, is an ethical theory which usually is taken to be a rival to consequentialism, Kantianism, and other theories on the long menu of options in normative ethics. A virtue ethics will naturally have a lot to say about virtue theory. But a Kantian can (and often does) offer a virtue theory, as well. Same with a consequentialist, a Rossian pluralist, and the like. There are questions about moral relativism which arise for virtue ethics, and some prominent virtue ethicists have flirted with relativism, but such matters will be put aside for the purposes of this chapter.[4]

Second, the virtues show up in different kinds of normative discussions. There are the epistemic virtues, for instance. There are also civic virtues, aesthetic virtues, prudential virtues, religious virtues, and so forth. My focus will be just on the moral virtues, although much of the discussion that follows could be carried over to these other normative realms.

The third preliminary point is that if moral relativism is true about *all* of morality, about all the moral truths, then there is not very much that is interesting to say about relativism and virtue *specifically*. Claims about moral virtues would be true or false in the same relative way that claims about other moral properties would be. Whatever motivation there would be to accept meta-ethical MR across the board about all of morality, would presumably apply to being a relativist about moral virtue. So in this chapter I will not consider the positive arguments offered for MR which apply generically to all the kinds of moral properties. Nor will I consider issues which arise specifically about MR and deontological moral properties like *morally ought* and *should*, as well as axiological moral properties like *moral goodness* and *value*.

With this stage-setting in place, what is my focus going to be? It will be on the question of what concerns arise with respect to *moral virtue specifically* that could lend some plausibility to a moral relativist approach to moral virtue.[5] In the next section of the chapter, I will consider three such concerns.

Before we get to those, let me end this section by setting aside three red herrings. These are issues about virtue which might have the superficial appearance of lending support to MR. But upon closer inspection, it seems clear that by themselves they do not:

- Many virtue theorists argue that there are no abstract moral rules, decision procedures, algorithms, or the like which dictate when to exhibit a given virtue and when not to.
- Many virtue theorists accept the context sensitivity of the virtues. In other words, what is involved in acting from honesty or compassion depends a great deal on the specific details of the situation in which the person finds herself. How courage is expressed when someone is drowning in the ocean, for instance, is dependent upon variables like whether the courageous person can swim, how far from shore the drowning person is, whether there is evidence of a rip tide, and so forth.
- Some virtue theorists align their view with a form of moral particularism (for helpful discussion, see Stangl, 2008). Particularism gets formulated in various ways, but a common commitment is to rejecting the claim that any natural feature of an action invariantly contributes to the rightness or wrongness of the action. Rather, in one situation it may contribute to the action's rightness, whereas in another it may not. Moral principles pick out such invariant qualities, so particularists like Jonathan Dancy often question the existence of such principles. On his view, particularists claim that "the possibility of moral thought and judgement does not depend on the provision of a suitable supply of moral principles" (Dancy, 2004, p. 7).

Each of these commitments does not give any ground to the moral relativist.[6] Moral realists are by no means committed to the objective morality involving abstract moral rules, or being context-insensitive, or being invariant in the way specified by particularists. There are plenty of virtue theorists who accept one or more of these claims, while also accepting some form of moral realism (or one of the other positions in meta-ethics which rejects MR).

Three legitimate concerns

With these three issues put to one side, here are three different topics pertaining to the virtues where legitimate concerns could be raised by moral relativists. In each case, not enough has been said thus far to address the concerns, at least in my opinion.

First concern: the list of virtues

Different virtue theorists have come up with different lists of the virtues. Aristotle has his list; Aquinas, Hume, and Nietzsche each have different ones of their own. Even if they do not always claim that their lists are exhaustive, there are conflicts with respect to particular virtues. On some lists, humility is a virtue. On other lists, it is not, and may even be considered a vice.

The abundance of such lists of virtues may foster worries about relativism. As Martha Nussbaum writes,

> The relativist, looking at different societies, is impressed by the variety and the apparent non-comparability in the lists of virtues she encounters. Examining the different lists, and observing the complex connections between each list and a concrete form of life and a concrete history, she may well feel that any list of virtues must be simply a rejection of local traditions and values, and that, virtues being . . . concrete and closely tied to forms of life, there can in fact be no list of virtues that will serve as normative for all these varied societies.
>
> (Nussbaum, 1988, p. 34)

Of course, strictly speaking, the mere existence of diverse lists does not by itself entail that moral relativism about the virtues is correct. This is no more plausible than the existence of diverse views about the shape of the earth or the existence of God entailing relativism in those domains. Diverse views can easily co-exist with an objective truth about the matter.

Nevertheless, one might still worry that in the case of the virtues, the *best explanation* for this diverse set of lists is that there is no fact of the matter as to what is the objectively correct list. Instead, it might seem to be a matter of individual (or cultural) opinion. So while granting that there is no entailment at work here, the advocate of MR still could make a case for MR being the best explanation for the diversity of lists out there.

The challenge for the moral realist with respect to the virtues, then, is to justify there being an objectively correct list of the virtues.[7] Ideally, the realist would also make some progress in telling us what is on that objective list, too.

Some progress has been made in addressing this challenge.[8] A particular helpful contribution is Martha Nussbaum's classic paper, "Non-Relative Virtues: An Aristotelian Approach" (Nussbaum, 1988). Claiming to largely be explicating Aristotle's own approach, Nussbaum notes that we encounter different domains of life which, "figure in more or less any human life, and in which more or less any human being will have to make *some* choices rather than some other, and act in *some* way rather than some other" (Nussbaum, 1988, p. 35, emphasis hers). Examples include:

(1) Fear of important damages, especially death
(2) Bodily appetites and their pleasures
(3) Distribution of limited resources

(Nussbaum, 1988, p. 35)

Given such domains, we then get a "thin" account of each virtue, which will just be "whatever it is to be stably disposed to act appropriately in that sphere" (Nussbaum, 1988, p. 35). Hence there will be a virtue of acting well with respect to bodily appetites and their pleasures, although this approach won't deliver a "thick" account of what that virtue involves.

Nussbaum's approach provides a nice framework to the moral realist for thinking about the diversity of lists of the virtues. As she writes,

> People will of course disagree about what the appropriate ways of acting and reacting in fact *are*. But in that case, as Aristotle has set things up, they are arguing about the same thing, and advancing competing specifications of the same virtue. The reference of the virtue term in each case is fixed by the sphere of experience.
>
> (Nussbaum, 1988, p. 36, emphasis hers)

There is much more to be said about Nussbaum's approach. But given limitations of space, let me simply register my view that an advantage of her approach is that, if we can exhaustively canvass the domains of life, we may be able to justify there being a certain fixed number of virtues (for a similar observation, see Stangl, 2018, p. 518). Any list that is too long or too short can be criticised on this basis. On the flip side, an obvious limitation of this approach is that, even if it is sound, it would only deliver a thin account of each virtue. This takes us to the second relativist concern.

Second concern: the choice of thick virtues

Suppose we grant that there is a virtue corresponding to however many morally relevant domains there are in human life. At this point, a second relativist concern naturally materializes. Take a particular moral domain. There are often going to be multiple candidates among the character traits for being a virtue *with respect to that domain.*[9] Given this, why think there is any privileged or objectively correct choice among the candidates? Perhaps it is just a matter of individual or cultural decision.

For instance, consider the domain which pertains to evaluating one's own standing and self-worth. One candidate for the domain-appropriate virtue is humility. But another is magnanimity. And, if it is different from magnanimity, a third is pride. Different philosophers can be found giving different answers here. But is there really one correct answer?

The second challenge for the moral realist with respect to the virtues, then, is to justify there being one particular character trait which objectively is the virtue in each domain. Ideally, the realist would also make some progress in telling us what that virtue is.

What should the realist make of this challenge? With respect to some domains, I think the realist will not have much to worry about. My case study of honesty later in this chapter is one such example. Another is courage. Still another is justice.

Note that the realist can make this claim, while still acknowledging that there may be deep disagreement on how to *analyse* the virtue in question (which will be the third concern to be presented in a moment). Compassion, for instance, could be the virtue which objectively pertains to helping other people, even if there is debate about whether compassion requires, say, altruistic motivation.

But while there is not much to worry about for the realist with respect to some moral domains, it is also striking that attempts to come up with a *general procedure* to identify and justify particular character traits as virtues have largely failed. Arguably the most prominent proposal in the contemporary virtue literature was the naturalistic approach developed by Rosalind Hursthouse (1999); for other approaches, see Foot (2001); MacIntyre (1981, 1988). Unfortunately, Hursthouse's approach has faced a number of problems, as have the other attempts that have been developed in the literature (see Gowans, 2011; Stangl, 2018).

Third concern: substantive accounts of each particular virtue

Even granting that: (1) there are X-many moral domains for which there will be a moral virtue, and that (2) there is a correct choice of which character trait is the virtue in each of those domains, it is still the case that for any given virtue, there are multiple proposals that have been made for the correct account of that virtue.[10] Why think that there is any privileged or objectively correct choice?

For instance, to take the same example from before, compassion could be the virtue which objectively governs the domain of helping others. But there could be deep disagreement about whether there is an altruistic motivational requirement on compassion or not. Is one of these options objectively preferable?

So the third challenge for the moral realist is justifying there being a correct account of each virtue in each domain.[11] Ideally, the realist would also make some progress in telling us what those accounts are.

How forceful is this challenge for the realist? Of the three, it seems to be the least serious, *provided that* the realist has a compelling response to the first two challenges. In other words,

> *If* we do not accept moral relativism across the board from the outset,
>
> and
>
> *if* we accept that there is an objective list of the virtues,
>
> and
>
> *if* we accept that a given character trait really is one of those virtues,
>
> then
>
> relativism about different accounts of that virtue might not seem very plausible.[12]

Moral relativism is a metaphysical position. But if one already concedes that a given virtue-property like compassion exists objectively, then the existence of competing accounts of compassion seems like an epistemological matter. It may be very hard to tell which account is better, but that by itself does not do anything to favour moral relativism.

Matters are different, though, if one is not allowed to assume the three "if" claims stated previously. Then the diversity of competing accounts of a given virtue could be used by the relativist as support for being a relativist about the virtue itself. So it would seem that much of the discussion of this third concern will be parasitic on the results of the first and second concerns. They seem to be where the main action is.

Having said this, it is undeniable that much more work needs to be done in providing an account of each virtue individually. In some cases, there is almost nothing available in the recent philosophical literature about an important moral virtue. The virtue of generosity, for instance, has been ignored by philosophers, with fewer than five papers appearing in the past 50 years.[13] So too with the virtue of honesty, as we will see in the next section. Other virtues like graciousness, courage, and fortitude have not fared much better.

Hence, there needs to be a lot more progress made in articulating detailed accounts of a given virtue, followed by a cost-benefit assessment in order to try to see which one is the most plausible overall. As Nussbaum notes, "we can understand progress in ethics, like progress in scientific understanding, to be progress in finding the correct fuller specification of a virtue, isolated by its thin or 'nominal' definition" (Nussbaum, 1988, p. 37).

A case study: the virtue of honesty

To help illustrate the three challenges and explore them in more detail, I want to end the chapter with a case study of the virtue of honesty. While my discussion here will be brief due to limitations of space, much more detail can be found in my recent book manuscript on this virtue (Miller, forthcoming).

Recall that the first concern had to do with there being diverse lists of the moral virtues. Connecting this to Nussbaum's proposal, we need to identify the relevant domain for which honesty might serve as the virtue. On my view, that domain has to do with how we respond to the facts (as we see them). Stealing, lying, cheating, and misleading others are all ways of falling short in this domain, and they all have to do with not responding appropriately to the facts (for details, see Miller, forthcoming, Chapters 1–4).

So there will be a virtue corresponding to the moral domain of responding to the facts. That gives us a thin description of the virtue. From there, we have to confront the second concern of whether there is a particular thick virtue which can plausibly be said to objectively correspond to that domain. Fortunately, at least for this domain, there is an obvious response to this concern, namely the virtue of honesty. This does not seem very controversial. Overwhelmingly, honesty shows up on most people's list of the virtues. Indeed, it is arguably the most straightforward and least contested choice of a virtue for any moral domain.

That still leaves us with the third concern and the question of what to think when there are competing accounts of a virtue. Unfortunately in the case of honesty, it is not that there is a host of competing accounts out there in the philosophy literature. Rather, honesty is an example of a strikingly *neglected* virtue amongst philosophers. To quantify this neglect, there has not been a monograph on honesty in at least 50 years, and until the recent work of our team at Wake Forest, there have been almost no papers in leading peer review journals or edited volumes in philosophy in the same span of time. So it certainly counts as an example of a virtue for which we need a whole lot more work.

Having said this, my view is that if honesty is objectively the virtue pertaining to how we respond to the facts, then there is going to be such a thing as a better and worse account of honesty. Let me end this section with two illustrations of controversial claims about the nature of honesty for which I think there is a clearly superior view.

To help with this task, let me introduce the basic framework that I have used in other work to begin to unpack what is involved in being an honest person (Miller, 2017, forthcoming):

> (H1) – The virtue of honesty is the virtue of being disposed, centrally and reliably, to not intentionally distort the facts.

The "reliably" has to do with a virtue being consistent both across situations and across time. Hence, the honest person behaves honestly in the office, bar, and classroom, and not just occasionally but predictably over time.[14] The "intentionally" is meant to contrast with "accidentally". Someone who breaks the rules of a game by accident is not being dishonest, nor is someone who walks out of a store with an item in her pocketbook that happened to fall into it without her knowing. Finally, "distort" is meant to correspond roughly to "misrepresent". If my son wants to go outside to play and tells me that he finished his homework when he knows full well that he did not, he is intentionally distorting the facts.

Now a lot more needs to be said about this framework, and I do not expect readers to jump onboard with it from this brief description. My only purpose here is to simply introduce a concrete account to serve as a point of reference. It is by no means my final account; I end up introducing about 12 different modifications to it (Miller, forthcoming). But I do think it is the best starting point that I know of. If, though, the reader has a better proposal in mind, feel free to use it instead, as the main points I make do not hang on accepting (H1).

Let us return to the two controversial claims about honesty I alluded to earlier. The first has to do with whether honesty requires responsiveness to the facts as they actually are, or whether responsiveness to subjective representations of the facts is all that is required. Here is a case to illustrate:

> *Age of the World, Part One* – Jason genuinely believes on religious grounds that the universe is less than 50,000 years old. But he also knows that sharing this belief with his friends could make him a social outcast. So when asked, he tells his friends that the universe is billions of years old.

In this case, my intuition is that Jason does something dishonest, even though what he said is objectively true. Contrast this with a different version of the case:

> *Age of the World, Part Two* – Jason genuinely believes on religious grounds that the universe is less than 50,000 years old. He also knows that his friends will be accepting of him no matter what he shares about his views. So when asked, he tells his friends that the universe is less than 50,000 years old.

Now I am strongly inclined to say that Jason is being honest, even though both his belief and what he says to his friends are objectively false. As a result, I am strongly inclined to reject (H1) and replace it with:

> (H2) – The virtue of honesty is the virtue of being disposed, centrally and reliably, to not intentionally distort the facts as the agent sees them.

The key point here is not about the entirety of (H1) or (H2), but rather just the clause about the actual facts versus the facts as the agent sees them.

So we have two competing claims about the nature of honesty. One could think that there is a relativist challenge looming, but I don't feel any such pressure in this case. It seems clear to me that (H1) is a second-rate proposal as compared to (H2). And if we are already onboard with the objective existence of the virtue in the first place (in response to the second concern from the previous section), then (H2) does a better job at clarifying at least one small part of what that virtue involves.

Here is the second example of a controversial claim about the nature of honesty. It has to do with whether the virtue pertains to cases in which it is plausible to think that lying is morally permissible. The Nazi at the door case serves as a helpful illustration. If the Nazi is doing a routine pass through the neighbourhood looking for Jews, and you are hiding a Jewish family in your basement, the widely held view (certainly among my students!) is that it is morally obligatory to lie. Let's suppose this is right, despite attempts by Kant and some recent writers to argue that lying is always wrong (see, Griffiths, 2004; Tollefsen, 2014).

A question as far as the virtue of honesty is concerned becomes which of these two options is preferable:

(1) The virtue of honesty does not apply to certain cases, say giving false information to the Nazi in order to protect a Jewish family
(2) Lying in certain cases, such as to the Nazi in order to protect a Jewish family, is still a failure of honesty, but it is all things considered morally permissible

On the first option, the scope of honesty only pertains to the realm of the morally permissible. This would require a revision to (H2), along the following lines:

> (H3) – The virtue of honesty is the virtue of being disposed, centrally and reliably, to not intentionally distort the facts as the agent sees them, so long as it is not morally permissible to do so.

One way to develop this view is to resuscitate the old idea that lies require that the intended audience have a right to know the truth. Since the Nazi has no such right, the homeowner would not be lying and so thereby would not be doing something dishonest. There is a lot more to be said here, and it is worth stressing that this is only one way (and perhaps not even the most plausible way) of trying to elaborate on option (1).

Option (2) is more straightforward. The homeowner would still be lying to the Nazi, it would count as an instance of a dishonest action, and yet it would not be morally wrong but rather permissible, all things considered. How? One way would be by appealing to conflicting virtues, and arguing that considerations of compassion require protecting the Jewish family, and these considerations outweigh the relevant considerations in favour of truth-telling. But again, this is just one way to develop the proposal. The key thing to note is that with option (2), no revision would be needed to (H2).

So again, we have two competing proposals for thinking about an aspect of the virtue of honesty. And again, I would argue that there is a clearly preferable option, namely (2). I have made the case for this at length elsewhere, and won't revisit the details here beyond just making a few quick points (Miller, forthcoming). The first is that it seems intuitively obvious that the homeowner would be telling a lie to the Nazi if she were to report that there are no Jews in her house. Secondly, it seems intuitively clear that she would be exhibiting dishonesty, not as a person in general, but in her behaviour in this particular case. And finally, if intentionally distorting the facts is at the heart of what it is to fail to be honest (or so at least I say it is), then the homeowner is clearly intentionally distorting the facts.

Hence, I defend option (2) rather than (1). But I do not think this is merely a personal preference on my part, in a way that could lend support to a moral relativist approach to analyzing the virtue of honesty. Instead, it is my best attempt to elucidate another aspect of the nature of this objectively existing virtue.

The purpose of this section was to both further elucidate the three concerns about moral relativism and virtue, as well as suggest at least in a preliminary way that they did not seem very worrisome in the case of the virtue of honesty.[15] Hopefully, these two modest goals have been accomplished, while also making it clear that such a conclusion does not automatically hold for the other moral domains. We may need to go domain by domain, and see.

Final thoughts

One of the main takeaways from this largely programmatic paper is that discussions of moral relativism and virtue are still in early stages. Hardly anything has been said to support a realist perspective about the list of virtues and their various necessary and sufficient conditions. In addition, many virtues like generosity and honesty remain neglected. I hope future generations of philosophers interested in virtue will take up the following tasks:

(1) Provide many more accounts of the individual virtues
(2) Provide many more comparative assessments of differing accounts of an individual virtue
(3) Examine intuitions about individual virtues across cultures, to see to what extent there is agreement about how the comparative assessments turn out

This hope is well expressed by Rebecca Stangl:

> What we need . . . are rich philosophical accounts of the particular virtues and vices, drawing both from our own lived experience and cultural traditions, as well as cultural traditions differing from our own. Only once we have these accounts in hand can we determine whether the degree of cultural relativism that exists among them is problematic.
> (Stangl, 2018, p. 520)

In the meantime, while the moral realist seems to have the upper hand, both the moral relativist and the moral realist about the virtues should be cautious about declaring victory. That would be premature for now.[16]

Notes

1 I have devoted much more space to formulating moral relativism as a general view in meta-ethics, and to distinguishing these three kinds of MR, in Miller, 2002, 2009. See also Wong, 1991 and Gowans, 2011.

2 This version was inspired by Wong, 1991, pp. 442, 447. For a more moderate version, see Wong, 1991, p. 448.
3 I spell out what mind-independence involves in much greater detail in Miller, 2009. Note that the characterisation could have been stated in terms of moral facts rather than moral truths, but the difference will not matter for the purposes of this chapter.
4 For excellent overviews of virtue ethics and moral relativism, see Gowans, 2011 and Stangl, 2018. Among those who are often thought to have flirted with MR at some point in their writings are Alasdair MacIntyre, Bernard Williams, and Philippa Foot.
5 For a closely related discussion, see Stangl, 2018, although her focus is on virtue ethics as opposed to virtue theory.
6 The same is true for a moral pluralist position of the kind advocated by W. D. Ross, according to which there is no one supreme principle of moral obligation, but a host of non-reducible moral duties.
7 As Kristján Kristjánsson pointed out to me, this might be resisted if the realist also accepts the context-sensitivity of virtues, since that might allow for traits being virtues in one cultural context but not another. I am not sure what to say about this, but my initial inclination is to hold that context-sensitivity is meant to apply to the *expression* of the virtue (what form courage takes could vary significantly depending on the context), rather than to a trait's status as a virtue. But more needs to be said here.
8 One body of work that is relevant here is the taxonomy of 24 character strengths developed by positive psychologists using extensive and rigorous cross-cultural empirical investigation. For a review and critical engagement, see Miller, 2019.
9 Note that this is a very narrow version of descriptive MR.
10 For a similar observation, see Gowans, 2011, p. 404 and Stangl, 2018, p. 516. Again, this is a very narrow kind of descriptive MR.
11 This is more than just the idea of essentially contestable conceptions revolving around a core account of a virtue. The realist has to respond to the worry that the core itself is contested. Thanks to Kristján Kristjánsson for raising this distinction.
12 The appeal of moral relativism would then be traced back to the preceding debates about whether moral relativism is plausible for all of morality, whether there is an objective list of the virtues, and whether a given character trait deserves to be on that list.
13 For my attempt to change this, see Miller, 2018.
14 "Reliability" is not being used in a way that presupposes honesty. Thanks to Kristján Kristjánsson for flagging this.
15 To be fair, as Kristján Kristjánsson pointed out to me, I may have shown that some accounts of honesty are better than others, but that does not rule out there being still other accounts, such as (H13), (H14), and (H15), which are equally good. As far as this chapter is concerned, this is a fair point. I would only note that to show that some accounts of honesty are objectively better than others is by itself already incompatible with a relativist position about competing accounts of a virtue.
16 An earlier version of this chapter was delivered at the 2019 Jubilee Conference. I am grateful to the audience for feedback and to Kristján Kristjánsson and Catherine Darnell for very helpful comments. Work on this chapter was supported by a grant from the John Templeton Foundation. The opinions expressed here are those of the author and do not necessarily reflect the views of the Foundation.

References

Dancy, J. (2004). *Ethics without principles*. Oxford: Clarendon Press.

Foot, P. (2001). *Natural goodness*. Oxford: Clarendon Press.

Gowans, C. (2011). Virtue ethics and moral relativism. In S. Hales (Ed.), *A companion to relativism* (pp. 391–410). Oxford: Blackwell Publishing.

Griffiths, P. (2004). *Lying: An Augustinian theology of duplicity*. Grand Rapids: Brazos Press.

Hursthouse, R. (1999). *On virtue ethics*. Oxford: Oxford University Press.

MacIntyre, A. (1981). *After virtue*. Notre Dame: University of Notre Dame Press.

MacIntyre, A. (1988). *Whose justice? Which rationality?* Notre Dame: University of Notre Dame Press.

Miller, C. B. (2002). Rorty and moral relativism. *European Journal of Philosophy*, *10*, 354–374.

Miller, C. B. (2009). The conditions of moral realism. *The Journal of Philosophical Research*, *34*, 123–155.

Miller, C. B. (2011). Moral relativism and moral psychology. In S. Hales (Ed.), *A companion to relativism* (pp. 346–367). Oxford: Blackwell Publishing.

Miller, C. B. (2017). Honesty. In W. Sinnott-Armstrong & C. B. Miller (Eds.), *Moral psychology volume v: Virtue and character* (pp. 237–273). Cambridge: MIT Press.

Miller, C. B. (2018). Generosity: A preliminary account of a surprisingly neglected virtue. *Metaphilosophy*, *49*, 216–245.

Miller, C. B. (2019). Some philosophical concerns about how the VIA classifies character traits and the VIA-IS measures them. *Journal of Positive Psychology*, *14*, 6–19.

Miller, C. B. (forthcoming). *Honesty: The philosophy and psychology of a neglected virtue*. Oxford: Oxford University Press.

Nussbaum, M. (1988). Non-relative virtues: An Aristotelian approach. *Midwest Studies in Philosophy*, *13*, 32–53.

Snow, N. (2010). *Virtue as social intelligence: An empirically grounded theory*. New York: Routledge.

Stangl, R. (2008). A dilemma for particularist virtue ethics. *The Philosophical Quarterly*, *58*, 665–678.

Stangl, R. (2018). Cultural relativity and justification. In N. Snow (Ed.), *The Oxford handbook of virtue* (pp. 508–523). Oxford: Oxford University Press.

Tollefsen, C. (2014). *Lying and Christian ethics*. Cambridge: Cambridge University Press.

Wong, D. (1991). Relativism. In P. Singer (Ed.), *A companion to ethics* (pp. 442–450). Oxford: Blackwell Publishing.

2 Meeting the challenges to the universality of virtue

Jonathan Jacobs

Virtue-centred ethics might seem especially suitable for a relativist interpretation or a treatment of virtues as local rather than universal. That conceptions of virtues should be local to cultures, to traditions, to worlds of practice, to certain social and political forms, and so forth, might seem very plausible. It seems plain that in different cultures – and at different times even within a particular social world – there are different conceptions of what count as ethical virtues and different notions of virtuous practice. Perhaps conceptions of virtue reflect rival, or at least, strikingly different views of human nature, or at least human excellence. In recent decades, MacIntyre, for example, has helped revive the discussion of virtue, debates about the universality of virtue, and the relation between virtue and tradition. His *After Virtue* (1981) and *Whose Justice? Which Rationality?* (1988) in particular, has been prominent in those discussions. In the present chapter, the key concern is to sketch out some of the main outlines of a reply to claims that virtue-centred ethics should be understood in relativist terms. I argue that the case for relativism is actually quite weak, that relativism offers little illumination, and that many of the explanations of why virtues are not universal are impediments to a fuller understanding of virtues.

In addition, there is an important role for the character of civil society and the political culture in the argument I give. I do not attempt to show that there is one, fixed, unrevisable set of ethical virtues or that there cannot be important changes in conceptions of various virtues. However, relativism regarding virtues narrows moral anthropology and the understanding of ethical practice and life in needless, unhelpful ways. Finally, the universality of virtue is not a commitment with which we begin; it comes progressively into view as our understanding of virtue is deepened.

Granted, it is not difficult to imagine how ideas of human excellence, of moral requirements, and of what is choiceworthy and desirable, could be quite different in different places and times. Consider how conceptions of justice have changed – and continue to differ in different places, and are often disputed even within a context. Or, think of how the social roles and responsibilities of persons have changed and what that implies for how persons and their actions are evaluated. Expectations of people say, between the ages of 13 and 21, in contemporary economically advanced liberal democracies, are very different from what they were in 1750 or 1300. There is the oft-used contrast between Homeric virtues and virtues as understood by the Abrahamic monotheisms (which differ among themselves, as well). Confucian virtues may be similar in some ways to Aristotelian virtues, but the societies in which Confucius and Aristotle lived were quite different, and many of their values differed. And, surely, the virtues of persons in a very conservative, strongly religious culture are different from those in broadly secular societies in which change is a seemingly constant feature, and the corresponding notions of a well-lived life are very different.

In addition, there are metaethical considerations pointing to relativism, such as those highlighted by J. L. Mackie in his critique of ethical objectivity, and Gilbert Harman's arguments concerning what sorts of facts are needed for the explanation of human behaviour. Given the diversity of moral outlooks, the very broad range of conceptions of what count as virtues, issues of moral epistemology and the metaphysics of morals, it might seem that there are numerous, substantial reasons to endorse relativism. Also, arguments about the nature of moral motivation can point to relativism, if motivation is best understood in terms of affective or appetitive dispositions rather than cognitive ones. In that case, one might conclude that attitudes, responses, and desires, rather than understanding, shape and direct morally relevant action. Numerous different pathways of argumentation have led people to relativism. The view itself, however, can be characterized to a highly general level as follows: moral rightness and wrongness, and the correctness or mistakenness of moral judgment are relative to the norms and values people, at any given time and in any given place, happen to accept. That characterisation suggests relativism's detachment from objectivity and universality.

Moreover, important features of morality shared by many cultures could be explained in terms congenial to relativism. For example, moral rules against certain practices, such as murder, for example, are explainable in terms of needs and interests that almost any culture or community has, though there is no objective rightness to the prohibition of those practices. Every society has some system for allocating benefits and burdens among its members, purportedly with a view to the good of the society. Every society has practices concerning treatment of, regard for those who have died, but there is not some single, "right" way to do this. The relativist can acknowledge moral commonality but is likely to argue that: (a) to the extent that there are moral commonalities, they can be explained without postulating objective values. A given concept (for instance, generosity) might be understood to involve quite different types of conduct and attitudes; (b) differences between moralities can be adequately explained in terms of cultural, historical, and other differences; and (c) there is no method for ascertaining what the alleged objective values are. Any claims to objectivity would be 'inside' one or another culture and the value-claims would be shaped and informed by that culture.

Virtue-centred ethics might seem especially congenial to relativism, each offering support to the other. Nonetheless, virtue-centred ethics should be understood non-relativistically, non-locally, and it can provide reasons to reject relativism. Moreover, relativism actually affords no genuinely explanatory illumination, and it can be an impediment to enlarged, deeper ethical understanding. An important part of the view defended is that sound ethical virtues are not attainable in just any political or social culture, and this is because the relevant sort of moral education is not available in just any culture. Moral education has a crucial role in the argument presented. Virtues can be universal even if they are effectively encouraged and realized only in certain (very broad) types of circumstances. Indeed, it is empirically quite plausible that moral education leading to universally valid virtues occurs most effectively only in certain political/social cultures. This may seem a Whiggish interpretation, but it is not a triumphalist claim about "us", whoever "we" are. It is a claim about what conditions are conducive to virtue and those that are not. Not just any setting can promote virtue.

Humanity's understanding of its own nature, of human good, and how best to live is always changing and developing. It is susceptible to revision, explication, and elaboration, even if in some cultures or social worlds certainty is claimed. Recognizing the possibility of improved understanding helps us overcome the simplistic notion that there are clearly definable, distinct moralities, each with its own code, as it were. Socially, there are some such discrete moralities, but in another sense, they are all attempts at comprehending how human beings should live and many overlap in regard to core concerns.

James Laidlaw, an anthropologist of morality, notes the many neo-Humean accounts of morals "in which moral responses are explained by basic general requirements of social life and the nature of human emotions" (Laidlaw, 2014, p. 221). He goes on to say that these thinkers often regard the view as congenial to relativism in that "they claim that the facts of moral relativism – the evidence they take from anthropology of deep divergences between cultures in emotional responses and moral attitudes – support their naturalistic arguments" (Laidlaw, 2014, p. 221). He goes on to argue that:

> Unfortunately, these authors tend to cite headline conclusions proclaiming radical differences rather than examining good ethnographic accounts, or they might notice that the emotional response is not necessarily radically divergent at all. It is a mistake to assume that, because a practice is an established and approved "part of the culture," this means that people are required or expected to, or do, feel good about it.
>
> (Laidlaw, 2014, p. 221)

Exploring and explaining examples of this type of mistake, he concludes, "relativism is, in any case, an incoherent solution to a non-existent problem"

(Laidlaw, 2014, p. 25). Moreover, no group is in a position such that they have complete, unrevisable conceptions and realizations of virtues. The potential complexity of moral reality precludes that, despite subjective certainty. Even if people stop thinking about moral reality, its inexhaustible complexity remains available to reflection.

There are grounds to think that persons are most capable of addressing the complexity of moral reality through acquisition and exercise of virtues. The multiple, diverse aspects of human nature that are involved in the virtues enable us to be responsive in more discriminating, calibrated, and appropriate ways than other approaches to ethics afford. In addition, the most plausible and constructive approaches to moral education are approaches that encourage the development of virtues. In large part, this is because of how the virtues are responsive to the complex texture of moral reality; that reality can exhibit a multiplicity – a pluralism – of values without evidence of pluralism also being evidence for relativism. Such approaches have the best chance of succeeding at being genuinely morally educative even if encouragement of virtues is not an explicit aim of those approaches. There is a relation between what moral reality is like, on the one hand, and what moral education is like, on the other hand, such that moral education that cultivates the virtues is best able to facilitate comprehension of moral values, their complex interrelations, and responsiveness to them.

Moral education, civil society, and virtues

Moral education takes many different forms and even within a society, there might be numerous different strategies of moral education. Some are strict, rule-bound, and leave little scope for the development of individual judgment. This is sometimes the case in theocratic states, but there are plenty of secular versions of moral education that do not encourage the development of judgment. While in the People's Republic of China conditions are changing in many important respects, moral education – as is the case with just about everything – is managed by the state and its institutions. One of the essay questions on China's current university admission examination asks, "Drawing on your political knowledge, explain why the Communist Party should exercise leadership over the country's economy, armed forces, schools and all aspects of society" (30 June 2018, "The World's Most Important Exam is Flawed". *Economist*, 37). This is not theocratic, but it is rigorously doctrinaire. It is a good example of how a comprehensive, formulaic moral education can produce narrow, unimaginative attention and thought. The combination of ideology and codification can be seriously illiberal.

If moral education is structured by a fundamental principle (one that includes some specificity; not just "be good") or if it is oriented to instilling (unreflective) conformity, it is doubtful that persons will acquire a textured, articulate grasp of the complexity of moral matters. Granted, it might be part of the *point* of a conformist morality that there isn't much complexity, and that the one main principle or the few main principles are all that are needed. However, it is quite implausible to suggest that moral value figures in human beings' lives in ways that are always one-dimensional and uncomplicated. It is not untypical for a virtuous agent to recognise, for instance, that justice is required in the situation but also that considerations of compassion, honesty, and generosity also figure in the judgment of *what justice requires*. A multiplicity of moral considerations can make demands on our attention, receptivity, and judgment even in a single situation of ethical judgment and decision.

In fact, while the discussion of moral issues often focuses on one value at a time, as when we say that this is an example of acting justly, and that is an example of commitment and fidelity, and this other case is an instance of courage, in fact, diverse moral considerations are present in almost every situation. Moral life is not neatly compartmentalised into one-situation/one-value units, with the exception of a few complicated situations. When we are acquiring a grasp of moral concepts it is natural to focus on one at a time, and one type of consideration at a time. But that is to train our attention in certain ways, and to introduce us to specific types of considerations, and it is not because values are present just one at a time. As one matures, one finds that the intersection of multiple values is common and not exceptional. One might also find that there is not a fixed, systematic relation between values such that we can know in advance which values have priority over others.

One of the ways in which a narrow moral education can badly misrepresent moral reality is in the ways it will almost surely oversimplify moral anthropology. It will permit only a shallow conception of human beings' concerns, sensibility, and interests, and a misleadingly narrow appreciation of the forms of relation and interaction that are morally important to people. A moral education that is responsive to the texture of moral reality enables a person to acquire an informed, realistic appreciation of human beings. If one's morality is a rigid or overly simple code, it could be an obstacle to recognizing and appreciating a potentially rich diversity of aspects of human sensibility and concern. In doing that, it could also result in the person having stunted moral motivation and narrowness of moral concern. One concern about relativism is that it accepts forms of moral life as complete, and thus shuts down important avenues of enlarged understanding.

Imagine someone brought up in a very traditional society or in a society in which there are different castes or statuses and one is taught a sharply defined role or status-based morality. There might be a measure of wisdom in some of what one is taught but, insofar as the morality reflects a fixed hierarchy of statuses and roles and responsibilities appropriate to each of those, it will make it extremely difficult to recognize persons as full participants, as agents in a common moral world, and capable of learning from each other.

Participating in many different contexts of civil society can be a very important source of moral education. In those different contexts, one might encounter multiple forms of responsibility, different bases for praise, reprimand, and gratitude; one might acquire an understanding of different norms and the reasons for them, and learn how to give justifications, and what counts as an adequate rationale for one's actions. One can acquire an understanding of forgiveness, bases for censure, appropriate sanctions, and different contexts and forms of discretion and sources of self-respect. Participating in civil society can educate persons in the multiple senses of many morally relevant concepts or at least in the diversity of ways they apply in different contexts. That, in its own right, is a way of acquiring some understanding of moral anthropology, especially because a genuinely civil society is one in which persons are able to participate in many different contexts, and express themselves and their interests in a diversity of ways.

The kind of civil society referred to is not achievable in just any political order. In an illiberal, repressive political order, civil society does not develop into a richly textured sphere of voluntary activities and associations. In countries where the regimes are resolute about forbidding criticism, controlling the flow of information and ideas, and setting strict limits on voluntary interactions and associations, much of the point of such control is to prevent civil society from developing. What is unacceptable about it from the perspective of an illiberal, repressive regime is that in genuinely civil society, no one is *in charge*. Civil society's metabolism is largely shaped and animated by the voluntary activities of people in all sorts of aspects of their lives and they are not all jointly engaged in some overall common pursuit. Of course, institutions of the political order have various powers, but in a political order allowing extensive civil society, political leaders are not meant to be fully in charge; their powers are limited and they help maintain the rule of law and institutions necessary for civil society.

Participation in civil society can be morally educative in that people are thereby engaged in all sorts of relations, commitments, cooperative undertakings, contractual agreements, and planning, and in general, learning to negotiate a complex social, economic, political, cultural landscape via practical reasoning and many types of valuative judgment in diverse contexts. Given the variety of contexts and interactions, it can be an excellent way to develop ethical sensibility, awareness, perception, and judgment.

For example, the many different kinds of situations in which courage is required can enlarge one's understanding of courage and the diverse forms it can take. The courage required to tell someone the truth in a tense and difficult situation is different from the courage required to help drive away an assailant – but it is courage, just the same. Similarly, what justice requires cannot be fully specified by a principle or a code. Even if it is always a matter of seeing that persons receive what is due to them, many kinds of specific judgments come under that generalisation. What justice requires *might* be quite clear and explicit. Yet, even if certain rules specify clearly what is required in a given situation, there might be respects in which those rules narrow the moral focus artificially. Consider generosity; it might be shaped by convention in many respects, but is also open to original and diverse guiding insights and intentions.

The same is true of compassion, considerateness, patience, loyalty, and a host of other virtues. We noted that pluralism should not be confused with relativism; similarly, we should avoid mistaking various kinds of context-dependence as grounds for relativism. Whatever general characterisations of them might be accurate, they might not provide the specificity needed in actual moral judgments. We should not confuse the fact that values and relations between them can exhibit considerable context-dependence with a claim of relativism. Actually, the latter makes the relations between values much less interesting and illuminating. Moreover, as indicated previously, in just about any situation, numerous ethical considerations are relevant, and a crucial part of moral education is learning to make the integrative judgments that are responsive to the multiple values at issue. While certain virtues are especially significant – justice, honesty, empathy, and courage are plausible examples – it is a mistake to think of virtues only in terms of a short list of fundamental virtues. There are all sorts of ways that a person can be attentive to, and responsive to valuative considerations that reflect the person's character, patterns of judgment, and sense of what is important, what matters, and why.

What is often overlooked is that there are different forms of engagement with different kinds of ethical considerations. Christine Swanton refers to what she calls "modes of moral responsiveness (or acknowledgement)" (Swanton, 2003, p. 21). She writes:

> They include not only promoting or bringing about benefit or value, but also honouring value (roughly, not dirtying one's hands with respect to value, e.g. by not being unjust in promoting justice), honouring things such as (appropriate) rules, producing, appreciating, loving, respecting, creating, being receptive or open to, using or handling appropriate things in appropriate ways.
>
> (Swanton, 2003, p. 21)

It is not as though *the exercise of virtue* is just one thing in terms of mode of activity. The virtuous person is very likely to exhibit a wide range of types of responses, engagement, interaction, extending of oneself, restraining one's reaction, and so forth. This is one of the reasons why virtue-centred theory often acknowledges that ethics defies codification. Because responsiveness to ethical considerations involves many forms and there are countless ways that values can be co-present in situations, there is not a set pattern of virtuous responses. Moreover, "[t]here is no one circumstantially specified good life in which the virtues have to be exercised. Many different specific kinds of life, in different circumstances, can be lived virtuously" (Annas, 2011, p. 116).

Commitment to exercising practical reasoning in ways responding to ethical value needs to be open to the vast complexity of the ways ethical value figures in our lives. One's ability to "read" moral reality depends on experience of different kinds, facilitating the acquisition of concepts that can apply in a diversity of ways. Good moral judgment, decision, and action depend on an articulate grasp of the complex fabric of moral value. This is one of the chief reasons why participation in civil society is so important. It provides the contexts of moral education in which the many modes of engagement and the many forms of responsiveness can be learned. Relativism could needlessly separate and isolate contexts of judgment and action that should be understood as elements of overall moral understanding and practice.

It is through practical activity in multiple contexts that the agent can acquire a more informed, more articulate comprehension of good as one matures. One does not acquire virtues accidentally, yet neither does the aspiration to virtue require making virtue an object pursued in the way one searches for a lost item or even in the way one practices to become better at throwing strikes or playing the piano. Instead, the agent is habituated in ways that encourage and support moral learning through dispositions of attention and meeting expectations of acting well.

Again, in the contexts of civil society, there is broad scope for voluntariness to be exercised through which persons can acquire an understanding of the rich texture of moral concepts and requirements. One can enlarge understanding of how virtues jointly figure in a conception of leading an excellent life. Julia Annas writes:

> The different virtues appear to be focused on different values, but the virtuous life consists in the living of a life, not its circumstances, and so the diversity of ways of life and the values they focus on is not a barrier to the unification of the virtues, and thus an aspiration to a good life overall.
>
> (Annas, 2011, p. 116)

The good of a good life is not independent of morality, but moral comprehension and virtuous activity can be realized through differently led good lives and their diverse commitments.

Virtue, good, and relativism

I have argued that a person is better able to acquire an enlarged moral understanding by participating in a more rather than less civil society. This is because of the range of contexts and modes of judgment and reasoning a person can experience. I am not talking about especially difficult ethical challenges or weighty dilemmas but a variety of contexts, situations, roles, and responsibilities. Through participation in a diversity of contexts and engagement with various purposes and responsibilities a person can attain a more informed, articulate conception of good.

Annas emphasizes the important role of a commitment to good. We should not confuse pluralism in the sense of there being multiple sources of ethical value with pluralism in a sense that indicates fragmentation; the notion that there is no integrative conception of virtuous activity through which good is realized in the exercise of practical reason. On the view that she defends,

> although in exercising different virtues the virtuous person is directed towards different values, the different virtues can, and should in a good life, be unified by the person's living their life in accordance with a unified conception of goodness, one to be discovered and exercised by the person's practical reasoning, which unifies the acquisition and expression of the virtues. The virtues are, then, unified into an integrated commitment to goodness and are exercised in the living of a good, flourishing life.
> (Annas, 2011, p. 115)

Against pluralists who deny the point (Nietzsche, for example), she argues that virtue "involves a commitment to goodness *because* it is goodness" (Annas, 2011, p. 111). She notes that, "we should not expect all commitments to goodness to have the same form" (Annas, 2011, p. 111). A diversity of virtues and a diversity of conceptions of good can be accommodated without denying that a virtuous person has a distinctive commitment, exhibited regularly in the person's practical reasoning, to an overall, integrated conception of good. We can distinguish between: (1) a unified life in which a variety of ends, interests, and concerns are integrated by practical reason on the one hand; and (2) being committed to a pluralism of values such that there is no reason to believe they can be integratively realised through an overall commitment to good. In the latter, ethical judgment and ethical life are in a problematic predicament concerning coherence and integrity. The goodness to which the virtuous agent is committed is realisable through lives led in a great variety of ways and allowing for a high degree of individuality and cultural difference. However, the plurality of values is not a plurality of utterly independent and possibly conflicting values in the case of the virtuous agent. Virtue's concern with integrity of commitment can be very challenging, but it is integral to virtue.

A person does not acquire virtues by acting in conformity with a conception of good that is grasped prior to practical activity, nor can virtue "be assessed as a means to an already agreed-upon and determinately formed end" (Annas, 2011, p. 156). It is not a set, independently specifiable goal. In aspiring to virtue and in acquiring virtue, the agent is not getting better and better at achieving an end independent of virtue. Rather, the agent becomes a dispositionally virtuous person with an effective commitment to good. What Annas says of happiness is true also of virtue. She writes:

> Our conception of happiness becomes more determinate as we become more virtuous, as our character develops, we come to have a more

> integrated view of what is worth having and doing in life, and again this happens in a way which ceases to be explicit in our deliberations, but is still present in a way that can be recalled if required.
>
> (Annas, 2011, p. 163)

But the process of becoming virtuous chiefly involves a steadily enlarged and deepened grasp of what is worthwhile, admirable, and praiseworthy, along with the reliable disposition to *act* on that understanding. Such openness and responsiveness is another aspect relativism accommodates poorly.

Acquiring virtue and a sound conception of good jointly involve attention, awareness, receptivity, and an integrative appreciation of ethical considerations and deliberative excellence. Concern for good is not a distinct, independent preoccupation on the part of someone with an aspiration to virtue. The plasticity of virtue and the substantive diversity of good lives are not considerations pointing to relativism. Relativism cannot account for those features, except by carving moral reality and human good each into multiple, independent versions, and that does not facilitate moral education or knowledge of morality.

One's moral education inevitably is limited in various respects and will emphasise certain values and perspectives, leaving out others. Even in a diverse, dynamic civil society, it may be the case that many persons are raised to respect certain traditions and perspectives, or it could be that their experience is fairly narrow. Plus, one's domestic or occupational life might structure one's activities and experiences – including ethical experience – in quite specific ways. In any case, the fact that the society is open does not imply that all its members lead lives that are wide open to all sorts of different valuative perspectives or that they have interestingly broad ethical experience. It is likely that many will, but the openness of society could be, in large part, its accommodation of numerous traditions, each of which is fairly conservative and not broadly open to external influences. At the same time, there is the possibility of those communities and their traditions interacting and learning from each other. Often there is friction and distrust, but that is not the inevitable outcome of a diversity of traditions in a society.

The variety of ways of life and perspectives need not be an obstacle to the universality of virtues. On the contrary, it is in such conditions that one's moral perception and understanding can be tested, challenged, and revised in ways that disclose the universality of at least some fundamental virtues. If people are willing to interact on bases of respect and trust, that can enable them to recognise that their conception of, say, justice or compassion is too narrow or reflects partiality in an objectionable way. It can enlarge someone's appreciation of the forms that generosity and supportiveness can take, and it can highlight previously unnoticed ways that honesty and fidelity are important.

Granted, people very often are suspicious of difference or they are chauvinistic and more inclined to settle into a shallow version of "competitive culture" rather than being willing to try to understand affinities and differences between traditions, practices, and perspectives. But the main point here is this: in societies in which it is regarded as important that people should interact *as accountable agents* there will be much greater opportunity for moral education to occur in multiple contexts, and that can make a constructive difference to the understanding of anthropology – one that is informal but nonetheless highly important. Moral progress of the sort that is achieved through the interaction of traditions and perspectives often comes slowly, but it often comes, and that is the crucial point.

Sometimes it is claimed that the differences in ethical practices and views are a basis for relativism, for the view that the validity or soundness of moral beliefs, practices, and ideals is relative to whatever norms and values *people happen to accept* or happen to be regarded as authoritative in a given context. According to relativism, the rational justification of virtues need go no further than that. The present account rejects that form of localising of virtue.

There is an important, basic sense in which relativism is explanatorily useless. In the actual world of experience and action, we might find that some societies or cultures have moralities very unlike our own, but what is accomplished by interpreting that as evidence for relativism? What we probably mean by "moralities very unlike our own" is that, for example, their rules of justice, notions of generosity and hospitality, and what is admirable differ from our own, and that they express loyalty and friendship in different ways. This is not automatically evidence that their values are *so* different that they cannot be recognised as continuous with our own in some ways. Nor is it evidence that there is no basis for objective evaluation of them – or of our own values. Why would it be thought that there is no standpoint from which either of the ethical "worlds" could be evaluated and revised, with a view to genuine correction and not merely change?

Practices such as honour killings, mutilation punishments, and fully arranged marriages of persons at a very young age (or with the parties to be married being of very different ages) are parts of moralities very different from those that are prevalent in Western liberal democracies. The same is true in regard to slavery, public executions, and whole populations of human beings being regarded as vermin to be exterminated. In fact, however, all of those are elements of moral and cultural worlds with which our own has at least historical continuity, as well as being elements of contemporary moral worlds culturally distant from our own but co-habiting the planet with us. But, again, if there are good reasons why those are morally objectionable practices reflecting corrupt values, the fact that many people were or are committed to them does not make them any less objectionable and any less *really wrong*; and not just *from here*, but wrong, full stop. (They were wrong even when they were accepted practices. Why should that be an especially problematic fact? Don't human beings make serious mistakes about many things?)

The universality of virtue does not require that its exercise should be completely uniform across different contexts. We have already noted that within a particular social world, a virtue might take various forms in different contexts. Courage takes different forms in the context of a difficult personal admission and in the context of rescuing a person in a burning building. The conditions of people's lives can make a significant difference to the forms of exercise of the virtues. Yet, this does not imply that virtues are relative to cultural contexts or that there is no universal basis for the normative significance of a virtue.

What counts as the excellent exercise of a virtue *always* depends on the circumstances in which it is exercised. It is not as though there is a standard – excellent exercise of virtue V – that is independent of the realities of ethical life, and we then check how closely someone's conduct conforms to that universal standard.

This does not relativise the virtues. The social conditions in which a community lives might be morally objectionable, reflecting various significant injustices and other moral defects, the durability of which is attributable to a regime or group or political culture that has a tight hold on power. Suppose the regime has been able to impose a set of norms and values that discourage people from compassion, empathy, honesty, generosity, and openness, and people are indoctrinated to do the regime or that party's bidding; that is the ideal that is taught. There is no insurmountable epistemological obstacle to our recognising that in the repressive state the idiom of virtue has been corrupted and many people have acquired action-guiding dispositions reflecting that corrupt understanding.

Morally dubious or just plain wrong practices and norms can nonetheless be quite enduring. Where regimes manage to maintain repressive power, and where the culture endorses rationally questionable traditions, practices, and attitudes, people can become habituated to norms and values that might have little to recommend them, but they are largely secure against criticism. The fact that there are various objective goods for human beings does not assure that people will aspire to understand and realise them.

We do not start with an anthropology that tells us what are human goods and then face the practical problem of acquiring the virtues as means to the realisation of those goods. We face the practical challenge of figuring out how to lead our lives, and in the process, we attain conceptions of value, worthwhileness, and significance, and we – hopefully – aspire to have and exercise virtues. The virtues are not means to an independently specifiable end. Virtuous activity has a constitutive place in living well. It is not as though the anthropology tells us there is happiness or flourishing – which we characterize a certain way; and then there are the virtues – each of which makes some specifiable contribution to happiness or flourishing. Instead of that, there is the business of leading a life, and that involves developing states of character, many of which may be either virtuous or not, and it involves orienting oneself by various valuative commitments and guiding concerns.

Finally, we should consider the explanatory gains that would supposedly come from denying the universality of virtues. The chief explanatory concern should almost certainly be attaining an illuminating understanding of what people regard as virtues and why they so regard them. It is not surprising that in different cultures and in different eras, different characteristics are praised and admired. However, what is gained by denying universality in contrast, say, to seeking affinities, likeness, and what might be common rationales?

We can find commonalities despite the differences. The commonalities will be among some of the guiding values of those lives, the ways that they can be seen as lives in which a commitment to good is central. Those persons we can recognize as virtuous – even if much of what was praised and admired is different from what we praise and admire – are recognizable as such because they led their lives by exercises of practical reason that are not alien to us. In fact, there are many intelligible aspects. The same can be said of a good deal of ancient and medieval Chinese thought. For example, if we consider Legalism, Confucianism, and Taoism, and the differences between them, their implications, and what motivated them intellectually and politically, we find likenesses to positions and insights in ancient and medieval Western thought, likenesses that are neither forced nor artificial. There are important affinities in the anthropology and the conceptions of virtues and the debates between the various ethical/political positions.

The point is not that "people are the same everywhere"; there are sometimes deep and perhaps unbridgeable differences. There is a great deal that is practically impossible for any society, given its prevailing values and commitments. We cannot find ways to integrate a close counterpart of suttee, and as problematic as lethal injection is, it is hardly "our version" of drawing and quartering, if on no other grounds than the very different way the body is treated after the victim dies. Large-scale human sacrifice is probably a practice with very little likelihood of being part of Western culture for the foreseeable future even if there are plenty of examples of mass murder in its recent history. There are numerous, significant limits on what is practically possible for a society. That is not, in its own right, evidence for the localism of virtue. Maybe several of the things that are practically impossible should be; maybe they are deeply morally problematic. Or, maybe the society is so morally rotten that without radical change it is doubtful that it could come to a more sound moral understanding. However, isn't more to be gained by supposing intelligibility and then seeing what sort of depth and detail it can achieve, rather than starting with localism and then claiming to confirm its borders? That will always be a contestable matter, except perhaps with a practice such as sacrificial mass-killing or ritual cannibalism.

It might seem that practical impossibility can help us determine where the local limits are. But the issue is much more complicated than that. *Within* a given society, there can be a range of views about what is morally possible. People can regard each other as morally thoughtful, serious persons, even if they have various moral disagreements. They can recognise each other as having certain virtues, even if the modes of expressing them differ in some ways. Also, neither the universality nor the objectivity of moral considerations is sufficient for ensuring moral agreement. Why should we expect that there would be easy or swift convergence on what the virtues require or even on the list of what are the virtues? Of course some are, as it were, non-discretionary; the lives of practically reasoning beings such as us require that we have certain virtues in order to live well. But what exactly they require, and what other dispositions should be counted among the virtues, remain matters to be addressed by moral life.

Starting with the presumption that virtues are local, or concluding that they are on the basis of various facts about moral disagreement and practical impossibility, makes little helpful contribution to moral understanding. We do need to guard against assuming that our (whoever "we" are) understanding of a given virtue is the standard for it, and that it need not be open to revision. But if we acknowledge that our understanding is not the uniquely correct standard for the virtues, we will be in a much better position to enlarge and deepen and refine our understanding and our practice. If we allow limits of practical impossibility to, as it were, locate virtues at an untraversable distance that will impose needless limits on our ability to negotiate moral reality.

Perhaps we continue to be misled by a model of ethical theory in which there is the identification of a fundamental principle or criterion of right action, or there is a finite set of fundamental principles and anyone who engages in sound ethical reasoning does so on the basis of that basis, assuming that is the form universality takes. But that is a misrepresentation and simplification of ethical thought and ethical universality.

Ethical requirements cannot be codified, and there isn't some single value or even a small number of fundamental values that are addressed by all ethical judgments. There may be various contexts in which codification serves a genuine function, but they are not parts of an overall, comprehensive code. Virtue involves dispositions of receptivity and responsiveness informed by concepts, providing a rich vocabulary of practical reasoning. Dispositions uninformed by conceptual articulation fall short of virtue, as does a seeming grasp of relevant concepts without alert, discriminating sensibility. Someone could know the ethical code for a specific context, and even comply with it, without possessing virtue. For that person, the content of the code is just a set of rules to be followed, not an engagement with valuative reality on the basis of concern with good.

Our grasp of moral concepts is never complete, and our engagement with valuative reality can give us reason to revise our understanding. Aspiring to virtue will affect our motivational dispositions. Our conceptions of specific virtues can always be improved, and interaction with practices, judgments, and beliefs other than our own can help us enlarge and deepen our understanding – and our virtues. A virtue is open to revised articulation, adjustment in scope, and in application. The virtuous agent's conceptual grasp is not altogether rigidly fixed, even if it includes numerous prohibitions and requirements. The universality is not, so to speak, *given*; it is attained, by seeing how virtues apply in new contexts, different circumstances, and relations with each other.

References

Annas, J. (2011). *Intelligent virtue*. New York: Oxford University Press.

Harman, G. (1977). *The nature of morality: An introduction to ethics*. New York: Oxford University Press.

Laidlaw, J. (2014). *The subject of virtue*. Cambridge: Cambridge University Press.

MacIntyre, A. (1981). *After virtue: A study in moral theory*. Notre Dame: University of Notre Dame Press.

MacIntyre, A. (1988). *Whose justice? Which rationality?* Notre Dame: University of Notre Dame Press.

Mackie, J. L. (1977). *Ethics: Inventing right and wrong*. Harmondsworth: Penguin Books.

Swanton, C. (2003). *Virtue ethics: A pluralistic view*. New York: Oxford University Press.

3 Socratic piety and the universality of virtue

Randall Curren

Introduction

In the landscape of ancient Greek ethics, there are two conspicuous manifestations of commitment to the idea that virtue is universal or the same everywhere and for everyone. The first is Plato's response to the relativism of the sophist Protagoras, who held that justice and virtue are local conventions rather than universal and natural. The defence of natural justice and universality of virtue is a central task of Plato's *Republic*, continues in his derivation of an elaborate system of natural law in the *Laws*, and reappears in Aristotle's conception of ethical science (Curren, 2019). In all of the relevant works, there is a close relationship between natural justice and the whole of virtue; in Aristotle's terms, just laws communicate truths about living well and prescribe "the whole of virtue toward others" (Barnes, 1984, p. 1783 [*Nicomachean Ethics* 1129b25–26]). A second conspicuous example of commitment to the universality of virtue in ancient Greek ethics is Stoic natural law, which reappears transformed in Kant's natural law constructivism (Westphal, 2016a, 2016b). Socrates has meanwhile been widely conceived as entertaining a eudaimonistic form of virtue ethics, "centered on the question of what happiness is and how best to achieve it" (Annas, 1999, p. 37). It is characteristic of orthodox virtue ethics that it excludes any role for moral law, yet Plato and the early Stoics took themselves to be developing Socratic ideas, and "it is hardly an exaggeration to place the origins of natural law theory . . . firmly within the framework of the Socratic movement" (DeFilippo & Mitsis, 1994, p. 271).

The thesis I will advance is that the moral and theological commitments attributable to Socrates on the basis of Plato's dialogues and Xenophon's *Memorabilia* suggest an understanding of virtue informed by natural law and a related understanding of piety and *fidelity to reason*. Piety and natural law give virtue a defining content that is the same everywhere for all rational beings. Moreover, it is the foundational role of piety in Socratic virtue that unites virtue and moral law in one coherent package. No account of an ethical theory in the Socratic tradition can be complete without addressing this coherence, and it bears emphasising that this includes Aristotle's – the orthodoxies of contemporary virtue ethics notwithstanding. Nor can character education productively ignore the significance of rules in children's lives and related questions about the nature and role of just communities in character formation (Curren, 2016, 2020).

The motivational core of piety is proper regard for what is divine, and I will argue that in Socratic theology, what is divine is the rational psyche or intellect (*nous* or *to logistikon*) and its intellectual products (*logos*). Socrates purifies the gods into creatures of unalloyed rational intellect (Vlastos, 1991; Janko, 2009), holds that human beings are part beast and part god (rational intellect), and endorses the traditional Greek view that true kings administer laws that codify divine wisdom communicated by Zeus or Apollo (Lloyd-Jones, 1983, p. 6). While committed to honouring the deliverances of his own reasoning, Socrates also honours laws and signs he has reason to think are emanations of divine intellect or *logos* more perfect than his own limited wisdom. Proper regard for what is divine will consequently manifest itself not only in reverence for gods and the wisdom they dispense, whether through universal laws or specific signs, but also in respect for the reason in oneself and other human beings, and for the deliverances of one's own best reasoning. I take all of this to yield the view that *virtue involves a proper valuing of rational beings, both perfect and imperfect, and wisdom, expressed both in universal laws and in judgments and signs tailored to specific circumstances*. Proper regard for rational beings and demands of reason is the defining core of Socratic virtue.

John Cooper writes that:

> On the Stoic theory virtuous persons always act, first, for the sake of their own happiness . . . second, choosing virtuous acts for their own sake, as parts of their happiness; and, third, choosing them as conforming to the universal law, that is, to the will of Zeus. . . . The Stoics, like other Greek philosophers of their time and before, equate honoring and following Zeus with honoring and following reason. In that sense, as with Kant, this universal law is our law qua rational beings for ourselves, and not only Zeus's for us.
>
> (Cooper, 1996, pp. 262–263)

Cooper goes on to explain that this universal law that is our law qua rational beings for ourselves is much wider than moral law, and that the universal reason it embodies is both common to all human beings and "more fundamentally, the single reason that governs the unified world as a whole" (p. 278). As such, it provides no scope for lesser rational beings determining their own ends or claiming the rights of autonomy that Kant and other moderns accord them. The transformation of natural law in the modern period involves, first, an abandonment of the idea of unified rational governance or law written into nature and, second, a shift to a constructivist mode of justification. Moral laws were justified by Aristotle and other Greeks as demonstrable truths analogous to theorems of geometry, but by Kant and other modern thinkers as self-evident demands of practical reason (side-stepping the question of whether they also express truths) (Curren, 2019; Westphal, 2016a).

Understanding Socrates' formative role in the natural law tradition is important to understanding not only Socratic ethics, but the intermingling of virtue and natural law in Plato's and Aristotle's ethics. I shall begin with a brief summary of Joseph DeFilippo and Philip Mitsis' account of the Socratic materials on which the Stoics built, consider the nature of piety and divine status of intellect in Socratic thought, show how the ethical commitments attributed to Socrates in the *Apology* and *Crito* can be explained as manifestations of a unifying ethic of *fidelity to reason*, and close with some observations about Socratic ethics and character education.

The Socratic roots of Stoic natural law

DeFilippo and Mitsis hold that the Socrates portrayed by Plato and Xenophon "never explicitly formulates a theory of natural law. But the presence in Socratic texts of the basic components of such a theory enabled the Stoics to construct their own account on a recognizably Socratic foundation" (DeFilippo & Mitsis, 1994, p. 255). They rely on the first book of Xenophon's *Memorabilia*, where Socrates is presented as arguing that the existence of human reason (*nous*) can only be explained by reference to a rational power in nature analogous to the material elements (earth, air, water) from which human bodies are constructed (pp. 255–257, relying on *Mem.* 1.4.8–9). This suggests a view according to which each person's rationality is "a portion of rationality in the world at large", a portion of divine rationality, that is. It also suggests that human beings are products of a rational power that governs nature much as the human psyche governs the body; hence, the idea that divine rationality orders nature (p. 257). The context is an attempt by Socrates to persuade the atheist Aristodemus that piety is essential to happiness, so the larger picture that emerges is one of Socrates both affirming the ethical importance of piety and associating it with regard for the divine rationality of which our own rationality is a part. DeFilippo and Mitsis note the evidence that both early and later Stoic philosophers relied on the presentation of Socrates' views in *Memorabilia* 1.4, in developing and defending their own position on the divine rational ordering of nature (pp. 260–265).

With regard to divine moral principles, DeFilippo and Mitsis observe that a divine ordering of nature as such does not entail the claim that "nature provides guidance to individuals in the form of moral laws or rules" (p. 265). However, they argue that Plato's *Crito* and *Phaedo*, and Xenophon's *Memorabilia* 4.4, provided the Stoics with Socratic starting points for this aspect of natural law theory as well. *Memorabilia* 4.4.19–21 finds Socrates arguing that there are unwritten laws, established by gods and observed everywhere, that direct human conduct and are backed by inescapable sanctions:

> "I think that the gods established these laws for men; for among all men, the primary law is to honor the gods," he [Socrates] replied.
> "So, then, it is also a law everywhere to honor one's parents?"
> "Yes," he replied.
> (4.4.20; transl. DeFilippo & Mitsis, 1994, pp. 266–267)

These are categorical imperatives of morality applicable to all human beings, everywhere and at all times, and Socrates is portrayed as committed to such imperatives in Plato's dialogues (e.g., *Cr.* 51a–b), no less than in Xenophon's *Memorabilia*.

Following David Sedley (1993), DeFilippo and Mitsis focus on two matters in the *Crito* and *Phaedo*. One is the fact that Socrates reasons in the *Crito* from a moral principle from which he will not waver, that one should never commit an injustice, even in return for an injustice (*Cr.* 49b–c). Such reasoning from overriding principles is congenial to the standpoint of Stoic natural law, and evidently central to Socrates' conception of virtue. A second notable starting point for the Stoics, identified by Sedley and endorsed by DeFilippo and Mitsis, is Plato's portrayal in the *Crito* and *Phaedo* of Socrates' calm acceptance of his impending death. The Stoics evidently accepted this as a "model of the sage's affirmation of the rational and providential order of nature" (DeFilippo & Mitsis, 1994, p. 268), communicated through divine signs, such as the dream Socrates evidently took to signify his death two days hence (*Cri.* 44a–b). Taken together, Socrates' argument from principle and acknowledgement of a divine sign provided the Stoics with a model of "the absolute convergence, of moral principle and nature's divinely rational plan . . . a paradigm for the sage whose attitudes and actions most precisely embody the dictates of the *koinos nomos* [cosmic or natural law]" (DeFilippo & Mitsis, 1994, p. 270).

The nature of Socratic piety

The evidence adduced by DeFilippo, Mitsis, and Sedley suggests that Socrates regarded the rational psyche or intellect as divine, associated piety with respect for the divine rationality of which our own rationality is a part, and took fidelity to universal moral principles to be an aspect of virtue. These attributions require further attention, Did Socrates believe that the human rational psyche is literally divine? Did he have a conception of piety compatible with such a view of human nature? I will defend affirmative answers to these questions in this section.

Xenophon's *Memorabilia* may offer the most direct evidence concerning Socrates' position on whether human intellect is divine, but the idea that human beings are part beast and part god is a staple of the Socratic tradition. There are dialogues in which Plato has Socrates speak of becoming wise as becoming like God (e.g., *Phaedrus* 248), and of being virtuous as godlike or divine:

> God . . . is supremely just, and the thing most like him is the man who has become as just as it lies in human nature to be. . . . There are two patterns set up in reality. One is divine and supremely happy; the other has nothing of God in it, and is the pattern of deepest unhappiness.
> (*Theaetetus* 176c–e [Cooper, 1997, p. 195])[1]

Similarly, in the *Phaedrus*, it is the soul "that follows a god most closely, making itself most like that god", and nourishing itself on truth (248a–c [p. 526]) that secures the best fate for itself. In the *Timaeus*, what is said of a man who devotes himself "to the love of learning and to true wisdom" is that "caring for his divine part, as he does, . . . he must indeed be supremely happy" (90b–c [p. 1289]). In the *Alcibiades*, a dialogue that may have been authored by Plato, it is the part of the soul "in which knowing and thinking take place" that is said to be most divine, to resemble the divine, to be divine, and to be made good by wisdom (133b–c [p. 592]). Socrates is presented as saying that, "Someone who looked at [the thinking part of the soul] and grasped everything divine – god (*theon*) and intelligence – would have the best grasp of himself as well" (133c [p. 592]).[2] It is reasonable to ask whether Plato (or someone else) is presenting Socrates' own view in these passages or a "perfected" version of Socrates, but the concurring account of Xenophon suggests the former.

Did Socrates have a conception of piety consistent with regarding human intellect or rational psyche as literally divine? This might seem wildly at odds with the case for Socrates' piety presented in the *Euthyphro* and the *Apology*. In the former, Euthyphro impiously attributes uniquely divine powers to himself as he presumes and fails to define the nature of piety; while in the latter, Socrates piously insists that his so-called wisdom is no more than awareness that he lacks the godlike knowledge others falsely claim. "The god is wise", and "human wisdom is worth little or nothing" by comparison, says Socrates (*Ap.* 23a [Cooper, 1997, p. 22]; cf. *Hp. Ma.* 289b), but wisdom is "the greatest good" (*Cri.* 44d [p. 39]), and those who wrongly believe themselves wise are guilty of "the most blameworthy ignorance" (*Ap.* 21b–e, 29a–b [pp. 21, 27]). Such ignorance is blameworthy, like Euthyphro's, because it fails to acknowledge the immense gulf that lies between themselves and a god, and thus fails to love and strive toward a wisdom measured not by man, but by god. Julia Annas re-describes Socrates' disavowal of wisdom as a pedagogically motivated disavowal of "ideas of his own" (Annas, 1999, p. 21), but Richard Kraut's interpretation is truer to the text and more in harmony with the theological dimensions of Socratic ethics. What Socrates denies is not that he has true ethical beliefs but that he has the definition-grounded moral *knowledge* or wisdom that gods alone have (Kraut, 1984, pp. 230–231). Socrates' pious modesty is characteristic of Athenian religious tradition (McPherran, 1996), but it is consistent with the view that human intellect is divine, if only a portion of Zeus or Apollo's.

The *Euthyphro* also poses a puzzle about the sense in which the activities for which Socrates must defend himself against the charge of impiety can qualify as pious. As Gregory Vlastos says in commenting on the *Euthyphro*, Socrates assumes:

> that what piety *is* depends no more on what they [the gods], or anyone else, feel about it, than does the nature of fire depend on what anyone . . . happens to think that fire is. Piety, and by the same token, every other virtue, has an essence of its own which is as normative for the gods as it is for us.
>
> (Vlastos, 1991, p. 165)

This being the case, there must be something about the nature of the activity for which Socrates stands accused that qualifies it as pious – something that would justify the gods in regarding it as pious independent of their own desire or volition. It must be an activity that expresses in a fitting way a reverence or respect for something divine.

Vlastos and Mark McPherran have focused their analyses of Socratic piety on Socrates' conception of himself as a servant of god in his role as a street-philosopher, and they hold that the logic of this is that he conceives his attempts to persuade others to engage in moral self-examination as pious service to the gods because it promotes human happiness in a way that the gods would themselves if they could (Vlastos, 1991, pp. 173, 175, 177; McPherran, 1996, p. 80).[3] This is plausible, given Socrates' conviction that virtue is a prerequisite for happiness (e.g., *Ap.* 30b), but there are two problems with this interpretation. The first problem is that what Socrates says in the *Apology* is that his service to both god and the city is "nothing but persuading both young and old among you not to care for your body or your wealth in preference to or as strongly as for the best possible state of your soul" (*Ap.* 30a–b [Cooper, 1997, p. 28]; cf. *Phaedo* 64e). The goal of his persuasion is for people to care about being "as good and as wise as possible" or *value* their rational soul and it being as wise or godlike as possible (36c [p. 32]). The second problem is that although promoting happiness is something the gods would have independent reason to approve, approving it does not entail having reason to regard it as *pious* independently of their own desire or volition. How would promoting happiness as such, or even virtuous happiness, express in a fitting way a reverence or respect for something divine? Both of these problems are solved if Socrates regarded the rational psyche as divine and as strengthened by virtue and fulfilled in wisdom. Showing respect for the divine element in human nature and encouraging others to do so would qualify as *substantively* pious.

What one's own reason tells one in this regard should coincide with guidance emanating from a god's perfect reason, through signs or through natural law pertaining to piety. Given this and Socrates' references to heeding the oracle, it is reasonable to think that his conception of piety involves, in relation to human beings, both a *patient-regarding* substantive aspect, concerning regard for the reason in oneself and others, and a *directive-regarding* aspect concerning fidelity to deliverances of perfect reason or divine *logos*. With respect to the latter directive-regarding element, Panos Dimas has helpfully argued that the *Euthyphro* does not fully repudiate attempts to "account for piety in terms of the gods' wishes" (Dimas, 2006, p. 7). He concedes that by Socrates' lights, a person could not be pious "by being attentive to the gods' wishes only", but he argues that Socrates believed the gods want human beings to be virtuous, ignoring the gods wishes would be impious, so being pious "requires that the agent is properly responsive to the virtues out of appreciation for their own intrinsic value and that he *also* does so for the reason that this would please the gods" (pp. 20, 22). While this proposal is less vulnerable to the first of my objections to Vlastos' and McPherran's interpretation, it is no less vulnerable to the second objection. Although appreciating virtues is something the gods would have independent reason to approve, approval would not entail having independent reason to regard it as *pious*.

My own proposal provides a coherent account of what makes Socrates' philosophy in the public square pious. It is *substantively* pious because it displays proper regard for the divine element in human nature. Because it is substantively pious, it is something reason and a god would demand. A person who engaged in this activity motivated by due regard for a god's wisdom would display piety both substantively and in a *directive-regarding* sense.

Demands of reason

If there are demands of reason associated with (pious) respect for persons as rational beings, what are they? What substance do they contribute to Socratic virtue?[4] Socrates models unwavering commitment to specific principles in the *Apology* and *Crito*, and these principles can be seen as manifestations of an underlying ethic of *fidelity to reason*. Imputing this ethic to Socrates explains many things, including an unnoticed symmetry in his treatment of the state's duty to put persuasion before force, and the citizen's duty to obey good laws voluntarily and "persuade or obey" the state's commands. The principles on display concern basic aspects of treating persons as rational beings, and they are of fundamental importance for the cultivation and exercise of virtue.

The state's duty to persuade

In pleading his own defense in the *Apology*, Socrates argues that it is absurd to suppose that he would have knowingly corrupted his associates, and inappropriate to punish him if he did so unknowingly. He reasons:

> If I corrupt them unwillingly, the law does not require you to bring people to court for such unwilling wrongdoings, but to . . . instruct them

> and exhort them; . . . but you bring me here, where the law requires one to bring those who are in need of punishment.
>
> (*Ap.* 26a [Cooper, 1997, pp. 24–25])

Socrates addresses this to his accuser, Meletus, but the jury would presumably be bound by the same principle and would be in no position to impose a punishment on one who has not been appropriately instructed. This implies that a city cannot establish a general claim to punish those who break its laws unless it makes a policy of putting instruction first by ensuring everyone subject to the laws is suitably educated. Public authorities have an inalienable responsibility to ensure that all children receive adequate character education, in other words (Curren, 2000, 2020).

Socrates allows that punishment is sometimes required, so he evidently assumes that in some instances, the burden of instruction will have been adequately borne. It is not clear how he would reconcile this conditional approval of punishment with his doctrine that no one does wrong willingly, but he may assume there are cases in which the futility of further educative efforts is evident and a city must protect itself through other means. A city must put instruction and persuasion before force, on the principle that instruction is the appropriate response to ignorance that has not proven to be impervious to instruction. Yet, to deny it the use of force in the face of unyielding ignorance and wrongdoing would be to deny it the means to defend itself and preserve the good order on which the well-being of its citizens depends.

This is somewhat speculative, but it reconciles Socrates' statements about punishment and ignorance in a way that conforms to the general pattern of a principle of putting rational persuasion and instruction before force and violence. This principle seems to be itself only one aspect of a more general principle of putting reason and wisdom first in every way, which seems to rest on the axiom that *the rational soul is the divine element in human nature and worthy of respect and care, since all things divine are worthy of respect and care*. This axiom, which I call the *principle of fidelity to reason*, suggests itself as the foundation of Socratic ethics (Curren, 2000). Showing proper regard for what is divine or best in human beings would evidently involve cultivating and relying on rationality and reasoning to the greatest extent one can. This would mean cultivating rationality both in oneself and in others, and relying on reasoning in action and in speech; in moving both oneself and others to act, and in judging what to say and what to believe of the opinions of others. Socrates' regard for reason is on display in all these ways in the *Apology* and *Crito*, and the strategy of his defence at trial is largely to explain that his fidelity to reason, annoying as it may be, is an expression of both piety and goodwill toward the city and its youth.

Applied to cities in their dealings with citizens, the *principle of fidelity to reason* generates corollaries pertaining to the *manner*, *aim*, and *substance* of rule. Thus, in moving citizens to act in some ways and not others, the *manner* in which a city should rule is as much as possible through truthful instruction and rational persuasion, and it should listen to the arguments offered in return by its citizens and endeavor to judge those arguments on their rational merits. In the *Crito*, the laws of Athens address Socrates and Crito, saying "we only propose things, we do not issue savage commands to do whatever we order" (52a [Cooper, 1997, pp. 45–46]). They speak repeatedly of persuading Socrates (53a, 54b, 54c), and they present him with philosophical arguments and instruct him to observe what follows from those arguments (51c). Their mode of persuasion is truthful and philosophically reasoned. In saying their commands are not savage (*agriôs*), these laws imply that they are reasonable, and they seem to intend this not simply as a point about their substance and aims, whose merits are invoked at 50d–e, but about their manner.

The *aim* of a city's laws and rational persuasion should be to produce not fidelity to commands and conventions as such, but fidelity to reason, and through it a grasp and pursuit of what is truly, and not just apparently, good. The city should aim at the highest good of its citizens, in short. This is essentially what Socrates asserts in suggesting at *Apology* 36b–37a that it would be appropriate for Athens to subsidize his philosophical examination of his fellow Athenians, and in saying at *Gorgias* 521d–e that his speeches make him a practitioner of the "true political craft" (Cooper, 1997, p. 564).

In its *substance*, the laws, commands, and verdicts by which rule is transacted should strive to embody reason, which is to say divine reason (*logos*) or laws (*nomoi*). Two passages where Socrates commits himself to the existence of divine laws are *Apology* 30d and *Crito* 54c. He says in the first of these passages that "I do not think it is permitted that a better man be harmed by a worse" (p. 28). In the latter, he has the laws of Athens speak to him of their "brothers, the laws of Hades (*adelphoi oi en Haidou nomoi*)", thereby identifying themselves as the laws of Zeus – those that would apply "in the air" or neither underground nor at sea. This suggests not only that Socrates believed in divine or natural laws that codify and communicate divine or perfect wisdom concerning how human beings can live well and harmoniously, but that he regarded such laws as communicated to at least some human lawgivers, as Greek tradition held.

The citizen's duty to persuade

Similarly, in its application to the conduct of citizens towards the city, the *principle of fidelity to reason* generates corollaries pertaining to the *substance* and *aim* of individual conduct, and to its *manner*, both in *compliance* and in *defiance* of law.

In the *substance* of their conduct, citizens should do what is right, just, or demanded by reason (*Ap.* 32a; *Cri.* 48c, 49a, etc.), and they should *aim* at the good (*Ap.* 29e–30a, *Cri.* 54b). To the extent that the laws of actual cities are not divinely inspired or grounded in systematic moral knowledge, those laws will sometimes demand acts that are just and sometimes demand acts that are not just. When they demand just acts, fidelity to reason demands that citizens comply in a voluntary manner, for in this case, what the law demands is what reason demands. To fail in this is to behave not in the *manner* of a citizen, but of a slave who is unable to govern himself in accordance with reason and must be forced to cooperate with it (*Cri.* 52d). On the other hand, when the laws demand unjust acts, what fidelity to reason demands in the manner of their defiance of law is that they make a reasonable effort to persuade the city that what it demands is unjust. Here again, instruction is the appropriate response to ignorance that has not proven impervious to instruction. Socrates reasons that in a good city, a citizen's consent to the laws and debt of gratitude for the benefits conferred by those laws creates an obligation to "persuade or obey" (*Cri.* 51b, 51b–c, 51e–52a), and that to fail in this obligation is to do violence to the city (51b–c). One way to avoid doing the city violence is to obey, but when obedience is precluded by the wrongness of what the law demands, the citizen must avoid doing the city violence by making a reasonable effort to persuade it that what it expects is wrong. If the city has already shown itself to be unmoved by reason, an acceptable course may be to simply ignore the law, as Socrates himself seems to have done by defying the unjust orders of the Thirty Oligarchs "not in words but in action" (*Ap.* 32c–d [Cooper, 1997, p. 30]). Otherwise, the citizen should attempt to instruct and persuade the city through truthful and rational means, but if he fails to convince it that its law is unjust, he must nevertheless do the right thing, even in defiance of its instructions (*Ap.* 29d, 32a–d, 37e–38a). Having put instruction and persuasion first as much as possible, his fidelity to reason does the city no violence.

The burden of instruction and persuasion that the city and citizen both face is a responsibility to encourage rationality and provide truthful instruction and reasons that a reasonable person would accept as sufficient. This is evident in the *Apology*, when Socrates explains that it is good, just, and pious to "teach and persuade" the jury, but not to subvert its impartial application of law through appeals to pity (35b–d). It is similarly evident in the *Crito*, in Socrates' explanation of why he and his friend must judge the arguments for his escape impartially, even as he awaits execution (*Cri.* 46b–48a). Judging the arguments impartially and heeding those that prove strongest upon examination is at the core of how citizens and cities should determine what to think, say, and do. To judge according to law and as piety demands is similarly to judge in a way that is reasoned and respectful of impartial demands of reason.

On this interpretation, the demands of justice governing citizens in their relation to cities are symmetrical with the demands of justice governing cities in their relations to citizens, and the common origin of these demands in the *principle of fidelity to reason* explains this symmetry. Principled defiance of bad law by citizens and punishment of wrongdoing by the city are both acceptable on Socratic principles, but both are subject to the constraints of fidelity to reason and the primary reliance on instruction and persuasion it entails. For the city's part, it is not a duty to endure injustice when reasonable efforts to instruct and persuade unjust individuals fail, for fidelity to reason entails striving to exhibit rationality ourselves and encourage it in others, and the creation and preservation of good order through good law is essential to doing that. Nor, for the citizen's part, is it a duty to undertake futile and potentially dangerous efforts to reason with rulers who have already demonstrated their disregard for reason, nor to persist beyond giving reasons that would move a reasonable person. By ascribing an ethic of *fidelity to reason* to Socrates, we can thus make coherent good sense of what might otherwise appear to be unrelated or conflicting passages in the *Apology* and *Crito*.

Plato constructed the *Euthyphro*, *Apology*, and *Crito* not only to defend Socrates against the charge of impiety, but to display him as the most virtuous of all Athenians – a model of virtue. So what is Socratic virtue, and what lessons does it offer for character education?

The character of Socratic ethics

Honouring the best in ourselves, the best in each other, and related demands of reason is the heart of Socratic virtue. If piety involves respect for divine wisdom (*logos*) and for the laws (*nomoi*) in which it is codified for the purpose of guiding human beings in living well together, then the virtue of piety is foundational to the content of other virtues. Respect for persons and for demands of reason will animate all the moral virtues, natural law will specify some parameters of what does and does not constitute respecting others in common forms of human transactions, and impartial applications of laws and principles will be fundamental to virtuous deliberation and action. This will be true even if virtuous people can often see what a situation calls for without deliberating. Virtue and natural law come together in the idea that lawgiving is first of all a mode of instruction in moral truths essential to virtue, and only secondarily a legitimating basis for the use of force in the public interest.

There is much in this account that would be disputed by those who read Socratic ethics as an egoistically eudaimonistic form of virtue ethics in which moral law and duty play no role, so I shall close with a few words about such readings. Orthodox virtue ethics staked its distinctiveness as a movement in contemporary moral theory on the quest for a moral decision procedure that was neither deontological – duty and rule guided – nor consequentialist. It claimed precedent in ancient Greek moral theory. There was little historical basis for this claim, but neither were there well-established interpretations that explained the role of natural law in Greek ethical theory. The idea that Socratic ethics is an egoistic form of eudaimonism owes much to the influential work of Gregory Vlastos and his former students, and their accounts of Socratic ethics have focused on the relationship between virtue and happiness, without much reference to demands of reason.

Vlastos argued that Socrates' ethics are grounded in a "Eudaemonist Axiom", according to which "happiness is desired by all human beings as the ultimate end (telos) of all their rational acts" (Vlastos, 1991, p. 203). From this axiom, Socrates allegedly derives a "Sovereignty of Virtue" principle, according to which, "Virtue being the sovereign good in our domain of value, its claim upon us is always final" (p. 211). Vlastos identifies a "Sufficiency Thesis", holding that virtue is necessary and sufficient for happiness, as the key to the derivation (pp. 216–217). The derivation is supposed to go as follows: (1) "happiness is man's good" and the object of all rational action (204n20); (2) virtue is necessary and sufficient for happiness, because it is necessary and sufficient for the psychic health on which happiness depends; and therefore (3), in order to attain our highest good, namely happiness, we should always do what is just or virtuous.

An obvious defect of this derivation is that it does not begin to make sense of Socrates' obvious regard for reason as categorically sovereign. It does not establish virtue as sovereign at all, but rather as instrumental to human desire. Note, above all, that the Eudaemonist Axiom is an assertion about what human beings *desire*, and that the derivation concerns what is instrumental to the satisfaction of desire. Of course, if human beings were constituted in such a way that they could not act out of concern for anything that is not conducive to their own happiness, then it would be essential to persuade them that virtue, justice, and valuing of things beyond themselves are essential to their happiness. Yet, the case for valuing those things – beginning with gods, persons, and reason – cannot be staked on the instrumental value of such valuing, because the result would not be virtue.

We see Socrates insisting in the *Crito* that he is categorically the kind of person who always acts as reason directs him to (*Cr.* 46b). Is his object in that case not to do what is reasonable or respect the demands of reason, choosing acts "as conforming to the universal law", as Cooper says in explaining the three-fold nature of what the virtuous person acts for, according to Stoicism? And are not virtuous acts chosen for their own sakes, not simply for the sake of one's own happiness? Is Socrates' restraint in not undermining the rational judgment of the jurymen in his trial not for the sake of their integrity and out of respect for them as rational beings? He insists it is. The attribution to Socrates of having egoistic and instrumental grounds for respecting reason and rational beings is implausible. What is far more plausible is that he was a reformer not unlike Hesiod and Solon in defending an impartial rule of law associated in Greek popular morality with an image of Zeus as caring about justice among human beings and laying down laws that codify truths about how human beings can live well together. The eudaimonistic basis of natural or universal moral law in this scheme is not the happiness of the individual actor called upon to live virtuously, but the happiness of everyone subject to and protected by the law. The associated demands of reason are impartial, and their worthiness of respect rests with their impartial responsiveness to the requirements of human happiness.

Universal demands of reason give Socratic virtue a measure of specificity in its universality, but Socrates' ethic of fidelity to reason weds this to a valuing of rational beings that also gives it universality and ensures that the valuing at the heart of virtue is not simply an honouring of moral law or demands of reason. Nor is the model Socrates presents one of mechanical applications of principle, but rather of fidelity to principles and diligence in reasoning one's way through the complexities of situations in which one must act. Socratic virtue does not fit the mold envisioned by contemporary virtue ethicists, but virtue it is. Understanding its intermingling of regard for persons and regard for demands of reason is helpful in understanding subsequent developments in the Socratic movement, including Aristotle's theory of virtue, and it should also be helpful to character educators in thinking about the role of rules in character education. Rules are pervasive in children's lives and play an inevitable role in defining what count as good habits for children to practise, while developing a nuanced understanding of the limitations of rules is arguably an important aspect of character development (Curren, 2016). The rules to which children are subject in schools define not just the habits they are expected to adopt, moreover, but the terms of their success, ability to satisfy their needs, and future prospects. Socratic ethics and its further development by Aristotle are not bad places to begin in assessing what would constitute a just school community favourable to sound character education (Curren, 2020).

Notes

1 The translations of Plato's works are those collected in Cooper (1997), except where deviations are marked by notes, transliterations, or reference to another source.
2 Hutchinson makes a gratuitous substitution of *thean* (vision) for *theon* (god) in his translation of this passage, in Cooper (1997). The sense of the unaltered text is clear enough, and is echoed in Aristotle, *NE* I.4 1096a25, where he says that "good" may be predicated of "whatness" (substances), naming God and the intellect as examples.
3 McPherran says there that Socrates' way of questioning people is pious "because it is productive of the virtuous happiness which good gods desire for us" (McPherran, 1996, p. 80).
4 This section relies on arguments developed more fully in Curren (2000), relying on Kraut (1984).

References

Annas, J. (1999). *Platonic ethics, old and new*. Ithaca, NY: Cornell University Press.
Barnes, J. (Ed.). (1984). *The complete works of Aristotle* (Vol. 2). Princeton: Princeton University Press.

Cooper, J. (1996). Eudaimonism, the appeal to nature, and "moral duty" in Stoicism. In S. Engstrom & J. Whiting (Eds.), *Aristotle, Kant, and the stoics* (pp. 261–284). Cambridge: Cambridge University Press.

Cooper, J. (Ed.). (1997). *Plato: Complete works*. Indianapolis: Hackett Publishing.

Curren, R. (2000). *Aristotle on the necessity of public education*. Lanham, MD: Rowman & Littlefield.

Curren, R. (2016). Aristotelian versus virtue ethical character education. *Journal of Moral Education*, *45*(4), 516–526.

Curren, R. (2019). Wisdom and the origins of moral knowledge. In E. Grimi (Ed.), *Virtue ethics: Retrospect and prospect* (pp. 67–80). Dordrecht: Springer.

Curren, R. (2020). Punishment and motivation in a just school community. *Theory and Research in Education*, *18*(1), 117–133.

DeFilippo, J. G., & Mitsis, P. T. (1994). Socrates and stoic natural law. In P. Vander Waerdt (Ed.), *The Socratic movement* (pp. 252–271). Ithaca, NY: Cornell University Press.

Dimas, P. (2006). Euthyphro's thesis revisited. *Phronesis*, *51*(1), 1–28.

Janko, R. (2009). Socrates the freethinker. In S. Ahbel-Rappe & R. Kamtekar (Eds.), *A companion to Socrates* (pp. 48–62). Malden, MA: Wiley-Blackwell.

Kraut, R. (1984). *Socrates and the state*. Princeton: Princeton University Press.

Lloyd-Jones, H. (1983). *The justice of Zeus*. Berkeley, CA: University of California Press.

McPherran, M. (1996). *The religion of Socrates*. University Park, PA: Penn State University Press.

Sedley, D. (1993). Chrisippus on psychophysical causality. In J. Brunschwig & M. Nussbaum (Eds.), *Passions and perceptions* (pp. 313–331). Cambridge: Cambridge University Press.

Vlastos, G. (1991). *Socrates, ironist and moral philosopher*. Ithaca, NY: Cornell University Press.

Westphal, K. R. (2016a). *How Hume and Kant reconstruct natural law: Justifying strict objectivity* without *debating moral realism*. Oxford: Clarendon Press.

Westphal, K. R. (2016b). Kant, Aristotle and our fidelity to reason. *Studi Kantiani*, *XXIX*, 109–128.

4 Evaluative attitudes as the motivational structures of ethical virtues

Jonathan Webber

Introduction

Are ethical virtues the same for everyone regardless of geographical, historical, and social context? Or are they keyed to the local environment in a way that means they vary across contexts? Ethical virtues are character traits that govern one's behaviour across all the situations of one's life. They are broader in application, therefore, than strictly professional virtues, which vary to some extent from one profession to another. The virtues required to be a good surgeon are not all the same as the virtues required to be a good teacher. Ethical virtues, by contrast, might seem universal, since the fundamental structures of human life are the same for everyone. We all undertake enterprises aimed at achieving goals, for example, our activities are temporally bounded, and our goals, identities, and even continued existence are fragile and vulnerable. The ways that these structures play out in our lives, however, depend on our geographical, historical, and social contexts. Does this mean that ethical virtues themselves are universal, reflecting the basic structures of the human condition, or local, reflecting the conditions in which those structures play out?

We will approach this question through the motivational structure of ethical virtues. Although ethical virtues are often likened to skills, because they are developed through practice and are not adequately captured in propositional descriptions, they differ from skills in one essential way. You can possess a skill but have no inclination to use it. But you cannot be honest and have no inclination to behave honestly. An ethical virtue essentially involves motivations to think and act in the relevant ways. What are the motivational structures of ethical virtues? This chapter argues that the motivational structure of each ethical virtue is a cluster of evaluative attitudes. In the first four sections, we will consider how best to model the findings of psychological research into evaluative attitudes. The fifth section will argue that the motivational structure of an ethical virtue consists in a cluster of evaluative attitudes, and that in many cases the relevant attitudes will depend on the specific challenges of the local physical and social environment. Therefore, many virtues will be local in their details. The final two sections identify two varieties of universality in the attitudes that compose ethical virtues. The attitudes composing ethical integrity do not refer to the environment at all, so ethical integrity is a genuinely universal virtue. One attitude of the virtue of honesty refers to environmental features that are themselves universal, so honesty includes at least one universal component.

Evaluative attitudes in social psychology

The study of attitudes has been central to social psychology since the discipline's inception almost 100 years ago. An attitude, in the sense in which this term is used in social psychology, is any positive or negative stance towards something. The idea of an attitude is therefore very broad in scope. You can have attitudes about highly abstract things, such as music or politics, about very specific things, such as your own home or the book you are currently reading, or about things at any level of generality in between. An attitude must be about something, which is known as the attitude's object, but there is no restriction on what that object can be. An attitude is usually described relating its object to a positive or negative valence. Accordingly, it is often modelled as composed of a representation of an object, a valence, and a relation between them. This chapter will argue for a different model of evaluative attitudes. We will read the usual description as committed only to attitudes being positive or negative stances towards their objects.

Social psychologists often describe an attitude's positive stance as a liking or valuing, a negative stance as a disliking or disvaluing. Indeed, they tend to treat liking and valuing as interchangeable terms, and likewise with disliking and disvaluing. But these terms are not the same: to like something is to enjoy it and to dislike it is to find it unpleasant, whereas to value something is to consider it a good thing and to disvalue something is to consider it a bad thing. You can dislike something that you value, as when you force yourself to do some unpleasant task that you consider important. And you can disvalue something that you like, as when you think that smoking a cigar would be very unwise even though you would enjoy it. We are going to concentrate solely on attitudes that are correctly described as valuing or disvaluing. These are genuinely *evaluative* attitudes. Merely liking strawberry jam is an attitude, but not an evaluative attitude. Considering democracy to be a good thing or thinking that lying is always wrong, by contrast, are evaluative attitudes.

Attitudes are measured in two different ways. The traditional method across the history of social psychology has been to ask people to report their attitudes linguistically, to infer their attitudes from their deliberative actions, and to compare the two. For example, someone might be asked whether they approve of a particular charity, then on another occasion asked to make a donation to that charity. These are known as explicit measures of attitudes, because the attitude is indicated by an explicit report or an explicit decision. Over the past two decades, attitudes have also been inferred from inattentive behaviour, such as body language, or decisions taken at very high speed. For example, your attitudes towards white people and black people might be inferred from your high-speed decisions in a computer game where you need to shoot all and only the characters carrying handguns, or from the amount of eye contact you make with someone while in a conversation with them. These are known as implicit measures, because they do not record explicit reports or decisions.

It seems that both explicit and implicit measures can detect evaluative attitudes, where these are defined simply as positive or negative evaluative stances towards objects. But we should not assume that they detect the same kinds of psychological states. It might be that implicit measures detect one variety of attitude, and explicit measures detect a different variety. Or it might even be more complicated than that. A full theory of evaluative attitudes will need to resolve this question and provide an integrated explanation of the two kinds of measure. But this is not necessary for the purpose of this chapter, which is to sketch how research into evaluative attitudes provides a model of ethical virtues that shows how the question of the universality or locality of each virtue is to be resolved. A full theory of ethical virtues in terms of evaluative attitudes will need to explain how it is possible to resist or reduce the negative influence of attitudes detected by implicit measures (Rees, 2016; Webber, 2016). But those details are not required for our purpose here. It is sufficient for this chapter to focus on attitudes detected through explicit measures.[1]

The structure of evaluative attitudes

We can model the findings of experiments employing explicit measures by distinguishing two dimensions of an attitude: its content and its strength. The content comprises the object, the valence of the stance towards that object, and the degree of that valence. If you strongly approve of democracy, for example, then the content of your attitude comprises its being about democracy and its being strongly positively valenced. If you consider the current voting system to be, on balance, a bad thing, then the content of your attitude is its being about the current voting system and its being mildly negatively valenced. This content is determined by the attitude's base, which is the set of beliefs and desires relevant to the attitude's object. For example, the base of your attitude towards democracy might include the belief that democracy is the best way to keep the peace, a desire that peace be kept, a belief that democracy best respects the autonomy of citizens, a desire that autonomy be respected, and so forth. The attitude base contains the beliefs and desires that you would readily report if you were asked to explain or justify your attitude. These mental states have affective aspects, as well as informational contents.

Exactly how is an attitude's content related to its base? On one model, the attitude is a separate mental state that is caused by its base: because you have various beliefs and desires relevant to democracy and other political systems, you have formed a particular attitude towards democracy (Tanesini, in preparation, §§ 3.1.1, 3.1.3). However, it would be simpler to consider the evaluative attitude to be nothing other than the set of mental states that compose its base: the attitude's content is determined by the attitude base simply because the attitude is the same thing as that base. An analogy with an image's degree of resolution can help us here. To say that you are in favour of democracy is to present your attitude in low resolution. But you can zoom in and present the details of the attitude, the beliefs and desires that compose it. Reporting that attitude base is a higher resolution presentation of the attitude itself. This way of understanding attitudes and their bases is not only simpler than understanding an attitude to be distinct from its base. It also has an explanatory advantage, as we will see in the section after next.

The second dimension of an attitude is its strength. This is not to be confused with the degree of valence, which is part of the content. An attitude's strength, in this sense, is the degree to which it is held, which means how strongly it is embedded into one's overall cognitive system. A better description would be "firmness", but "strength" is the standard term in social psychology. There are various measures used to assess attitude strength. Standard measures include asking subjects how certain the attitude is, how central it is to their self-image, how important it is to them, and how well it reflects their more general values. For example, if asked about your attitude towards democracy, you might report that you are very much in favour of it, that you are certain of this attitude, that you see yourself as a democrat, that this is highly important to you, and that it reflects your general valuing of peace and respect for intelligent disagreement; in which case, you have a strongly positive attitude towards democracy and this attitude is strongly (or firmly) held.

Why should we think that these four measures are all measuring a single dimension of the attitude? Why not think that they are measuring four different aspects of the attitude? One reason is that their aggregate score itself has a predictive value. Attitude psychologists have found that an attitude which scores highly on the aggregate of these measures is more persistent across time, is more resistant to change by persuasion, is more influential over cognition, and is more consistently expressed in behaviour than an attitude that does not score highly on the aggregate of these measures. In the next section, we will consider two classic experiments that indicate this predictive value. But the way these experiments contrast "strong" and "weak" attitudes should not mislead us into thinking that this is a binary distinction: all attitudes have some degree of strength, which varies gradually along a scale from very weak (very low aggregate score) to very strong (very high aggregate score).[2]

Attitude strength and consistency of expression: two experiments

The significance of attitude strength can be illustrated with two representative experiments from the extensive attitude psychology tradition. One of these is the cognitive weight loss programme experiment (Axsom & Cooper, 1985). Subjects were gathered from across the local community through a newspaper advertisement requesting volunteers for an experiment into possible methods of weight loss. Subjects selected were classed as significantly overweight according to life insurance company measures. They were assigned to one of three groups. The high-effort group had to attend the lab and perform strenuous and boring cognitive tasks four times across two weeks. The low-effort group had to attend at the same times and perform much easier and shorter versions of the same tasks. The control group attended and performed no tasks. Subjects were asked to keep a record of their eating across the period of the experiment. They were weighed at the start of the experiment, on each of the four times they attended, a week after the final task, six months after that, and finally another six months later.

One week after completing the tasks, subjects from the high-effort group had lost an average of 1.76 lbs. (0.8 kg) in weight, whereas the low-effort group had lost less than half of that, and there was no significant weight loss in the control group. But the more interesting result comes later: after six months, the high-effort group had lost an average of 8.55 lbs. (3.9 kg); neither the low-effort group nor the control group had lost any significant weight. This differential was almost entirely maintained a further six months later. How could a brief course of high-effort cognitive tasks cause significant and lasting weight loss? Since subjects volunteered for this experiment, they already held positive attitudes towards weight loss, but since they were significantly overweight, these had not translated into consistent behaviour. Only the high-effort subjects needed to persuade themselves regularly that their goal was worth the effort of continuing with their tasks. This repetition seems to have strengthened their positive attitude towards weight loss, making it more influential over their cognition and behaviour, leading them to adopt and maintain relevant behaviours such as dietary restrictions or better exercise (Cooper, 2007, p. 171).

A sophisticated experiment conducted more than 15 years later complements this reading of the cognitive weight loss experiment. Subjects were asked to complete a questionnaire about their attitudes towards 32 different objects, one of which was Greenpeace. For each attitude object, the questions included a measure of the attitude content and all four measures of attitude strength. Subjects returned to the lab a week later to participate in an unrelated experiment, for which they were paid in coins. On leaving this experiment, they had the opportunity to donate to Greenpeace and were subsequently asked to report their attitudes towards Greenpeace. This complex scenario was designed to conceal from the subjects that what was being tested was the consistency of their attitude towards Greenpeace. Attitudes that scored highly on the aggregate of the four measures of strength in the initial questionnaire were expressed a week later in response to the opportunity to donate to Greenpeace and in the subsequent explicit report of their attitude. By contrast, attitudes with a low aggregate score of strength did not correlate with their subjects' behaviour or reports a week later (Holland, Verplanken, & van Knippenberg, 2002).

This experiment supports the idea that stronger attitudes are more consistently expressed in behaviour and in explicit report. Weaker attitudes seem to change from one situation to the next, at least over the course of one week. This supports the idea that the cognitive weight loss programme worked by strengthening the positive attitude towards weight loss among the high-effort subjects. A stronger attitude in favour of weight loss would be more consistently expressed in positive weight loss behaviour and in explicit judgments about whether weight loss is an important goal at times of relevant decision-making. The weight loss experiment goes further in suggesting that an attitude is strengthened by expending significant effort in pursuit of the attitude object, presumably through the internal process of justifying that effort to oneself. These indications of the importance of attitude strength and of one process of attitude strengthening are central components of this chapter's claim that the motivational structures of ethical virtues consist in evaluative attitudes. Before turning to that claim, however, we need to add one further component.

Strength, stability, and composition

Ethical virtues are traditionally considered to be stable qualities of the person, character traits that persist through time and across different kinds of situations. Since one can develop the virtues, indeed be held responsible for doing so, they are not fixed dispositions of the individual. But they are considered to be ongoing states that explain the regular occurrences of the right kinds of thoughts, feelings, and actions in the appropriate situations. If ethical virtues as traditionally understood really do consist in evaluative attitudes, as this chapter maintains, then those evaluative attitudes must themselves be persistent mental states. One long-standing controversy in the social psychology of attitudes concerns precisely this question: are attitudes persisting mental states individuated by their objects and elicited in relevant situations, or are attitudes simply transient mental items that are constructed each time they are needed? The claim that ethical virtues consist in evaluative attitudes does not require a single answer to this question that covers all evaluative attitudes. Rather, it requires only that the relevant evaluative attitudes can be persistent states.

One of the experiments that we have already considered supports the idea that some attitudes are indeed persistent states, for a persistent mental state would explain the consistency of reports and behaviour that the

Greenpeace experiment found to be characteristic of strong attitudes. By contrast, the absence of a persistent state would explain the inconsistency found in weak attitudes. A detail of this experiment offers further support. Although a weakly held attitude towards Greenpeace did not correlate with behaviour a week later, that behaviour did correlate with the attitude report immediately following it. This indicates that these weakly held attitudes were constructed in response to their situations: the attitude constructed in the questionnaire situation is different from the attitude constructed in response to an opportunity to donate; when an attitude is reported immediately after a behavioural response to that opportunity, the behavioural response is one of the features of the situation that influences the attitude construction. The experimenters concluded that their study supports the hypothesis that strong attitudes are persistent mental states, whereas weak attitudes are temporary constructions (Holland et al., 2002, pp. 874–875).

How is this conclusion consistent with the aggregate scores of the measures of attitude strength forming a scale, rather than a binary distinction? Since all attitudes have some degree of strength, why would the attitudes in one part of the scale be persistent mental states while attitudes in the rest of the scale are not? This problem arises only if we see an attitude as a mental item that is distinct from the beliefs and desires that constitute its base. On that model, the attitude is a mental state comprising an object, a valence, and an association between the two, and the strength of the attitude is the strength of that association (Tanesini, in preparation, §§ 3.1.1, 3.1.3).[3] To accommodate the idea that strong attitudes are persistent states and weak ones are temporary constructions, this model would need to claim that a persistent attitude requires the association between object and valence to be over a certain threshold of strength, with attitudes below that threshold fading away soon after their construction and expression. But it does not seem to have the resources to explain why this should be the case.

If we consider an attitude to be identical with its base, however, then we can explain the results of the Greenpeace experiment without distinguishing between attitudes that are persisting mental states and attitudes that are temporary constructions, for this model allows an attitude's strength to be a function of the strengths of the associations between the mental states that comprise its base. If these items are very strongly associated with one another, then when any of them is elicited by the attitude object or some representation of it in the situation, the whole set will effectively be elicited together. In this case, the attitude base will function as though it were one single persisting mental state. The attitude will be a virtual persisting mental state. But when one's beliefs and desires about some object are only weakly associated with one another, the set that comprise an attitude towards that object on some occasion will depend on which of those beliefs and desires are elicited in that situation. This model therefore explains why the degree to which an attitude is consistently expressed in report and behaviour is determined by its overall strength.[4]

Evaluative attitudes and the motivational structures of ethical virtues

If we understand attitude strength in this way, then we can see a clear connection with ethical virtues as these have traditionally been understood. For on that traditional conception, an ethical virtue is an evaluative state that is consistently expressed in judgment and behaviour only if it is sufficiently firmly embedded. The most influential model of virtue development has been Aristotle's view that virtues become refined and strengthened through critical reflection on their requirements and through application in action. Evaluative attitudes consist in a set of beliefs and desires, which are subject to rational consideration, and which become more strongly associated with one another the more they are associated in thought in action. The stronger an attitude is, moreover, the more readily it comes to mind and motivates action, which fits with the idea that in some situations, the virtuous person simply sees the right thing to do and acts accordingly without needing to deliberate (Rees & Webber, 2014a, §§ 1–4).

We should not conclude that each ethical virtue is an evaluative attitude, but rather that the motivational structure of each ethical virtue consists in a cluster of related attitudes, perhaps bound together by associations reflecting the agent's understanding of the deep relations between them. The ethical virtue of honesty, for example, comprises the attitudes towards lying, cheating, and stealing that one ought to have, along with an understanding of why one ought to have them and the deep conceptual relations between them. Ethical virtues might also involve skills, such that the virtuous person pursues their goals skilfully rather than clumsily, but the motivational aspect of each ethical virtue is provided by attitudes. Our earlier metaphor of high-resolution and low-resolution images can be used again here. The language of virtues specifies complex motivational states in fairly low resolution. Zoom in for a higher-resolution image that presents the set of attitudes that comprises the virtue, with associations of varying strengths between these attitudes. Zoom in again for an even higher-resolution image that presents the cluster of associated beliefs and desires that comprise that attitude, with associations of varying strengths between them.

This model of the relation between ethical virtues and evaluative attitudes entails that the question of whether ethical virtues are universal or local can be answered at higher or lower degrees of resolution. The language of ethical virtues describes the formal structures of good thought, feeling, and behaviour, individuated by the domains of life to which they apply. Building on Martha Nussbaum's interpretation of Aristotle and drawing on twentieth-century existential phenomenology, Irene McMullin has recently argued that the relevant domains are the essential structures of human existence, so that patience is the virtue of dealing appropriately with our temporal boundedness, modesty is the virtue of dealing appropriately with the significance of our goals for our identity and self-esteem, and courage is the virtue of dealing appropriately with human vulnerability (McMullin, 2018; cf. Nussbaum, 1993). On this view, to say that these virtues are universal is not to say very much. For even if temporal boundedness, goal directedness, and the fragility of our lives and identities are universal features of the human condition, the specific challenges we face in relation to them will differ according to our geographical, historical, and social contexts.

In the terms of our model, this means that the set of evaluative attitudes that comprise an ethical virtue differs according to local conditions. For people who live in a war zone, the virtue of courage will require the right attitudes towards violence and the threat of violence. For people who live in the rainforest or on the savannah, courage will require the right attitudes towards the local wild animals and climate conditions. For those of us fortunate enough to live in more peaceable and secure environments, the most significant components of courage might be our attitudes towards the risks faced in our professional lives and personal relationships. In each case, these components will be unified by a general attitude towards the importance of one's own safety and well-being in relation to other values. This structure explains why the courageous person would act bravely in genuinely novel circumstances, but will do so more easily and readily in more familiar situations. Many ethical virtues will follow this pattern: universal when depicted in the low-resolution terminology of virtue, local when depicted in the higher-resolution terminology of attitudes. Some will become even more local when we zoom in further to the specific beliefs and desires that comprise the attitudes. In the next two sections, however, we will consider two different ways in which an ethical virtue's component attitudes might themselves be universal.

Ethical integrity as a fully universal virtue

An ethical virtue will not be fully universal when its component attitudes are keyed to contingent features of the local environment. One way in which they can be keyed to the local environment is by having features of that environment as their objects. We have already seen that the virtue of courage includes evaluative attitudes about contingent features of the environment, so that the universal specification in the language of virtue disguises the local specification in the language of attitudes, which is revealed when we zoom in to view courage in higher resolution. This demonstrates that for an ethical virtue to be fully universal, its component attitudes cannot have contingent features of the local environment as their objects. In this section, we will see that ethical integrity is a fully universal virtue precisely because it is fully specified, even at the highest resolution, without reference to anything that could vary between environments. In the next section, we will see that an attitude can be universal even if it does refer to a feature of the environment, so long as that feature is essential to the human condition.

Ethical integrity is a complex virtue. Consider first the simpler character traits of constancy of commitment and fidelity to one's commitments. Constancy is the quality of remaining resolute in some commitment despite significant pressures that might undermine it. Fidelity to a commitment is the quality of acting on that commitment despite significant pressures to do otherwise. In their thinnest forms, these traits simply consist in the strength (or firmness) of the attitudes composing that commitment. But where someone exhibits constancy or fidelity across a range of commitments, this can consist in an attitude that one ought to remain constant or faithful to one's considered commitments. For this to be a general attitude, it must not specify which commitments one should remain constant or faithful towards. These richer forms of constancy and fidelity are therefore fully universal traits. At no level of specification does either trait refer to anything that might change from one environment to the next. The traits, their component attitudes, and the beliefs and desires that comprise those attitudes are all about the importance of maintaining or acting on commitments generally (Rees & Webber, 2014b, §§ 3–4).

Are these traits virtuous? An insistence on retaining a commitment because it is one's own seems rather like the vice of intransigence. The failure to see that the mere fact of possessing a commitment is not a sufficient reason to act on that commitment seems like the vice of obstinacy. And yet there does seem to be something ethically admirable about the attitudes that underpin these qualities. We can explain this if we see constancy and fidelity as components of the broader trait of ethical integrity, which also includes a strong attitude in favour of getting one's commitments and behaviour right. This further component explains why ethical integrity can require one to revise or even abandon a long-held commitment. It also provides constancy and fidelity with their ethical value: if one's existing commitments are the products of a strong drive towards getting things right, then one should indeed respect them as repositories of previous ethical reasoning. This is compatible with recognising their fallibility. This further attitude, moreover, does not refer to anything that might vary between environments (Rees & Webber, 2014b, §§ 5–6).

There is one more component required for ethical integrity, since the question arises of how we should balance the demands of constancy and fidelity, on the one side, and the aim of getting things right, on the other side. How much should we respect our standing commitments? How much should we see them as open to revision by new reasoning? Answering these questions is itself an aspect of ethical integrity. The virtue of ethical integrity therefore requires an attitude towards revising one's existing commitments in the light of new reasoning. This needs to be strongly held, but its precise content should be determined through the exercise of that virtue (Rees & Webber, 2014b, §§ 6–7). Even so, this content is purely formal. Like this virtue's other components, this attitude should not vary according to which commitments one is concerned about maintaining or revising. Rather, for this complex character trait to really be the virtue of ethical integrity all of its component attitudes need to be purely general.[5] This is ultimately why this virtue is fully universal: at no degree of resolution do we see any components that might vary between environments.

Credibility and the virtue of honesty

Ethical integrity is fully universal because its component attitudes do not refer, either directly or indirectly, to features of the environment. But a virtue could be fully universal even if some components do refer to features of the environment: if those features are essential to the human condition, they will themselves be universal. One essential aspect of the human condition is that we require information, rely on one another for information, and are capable of deceiving one another. Honesty is the virtue of dealing appropriately with this membership in an epistemic community. Honesty is a complex virtue, involving the right attitudes towards cheating and stealing as well as towards all forms of deceptive speech, including lying, misleading, and bullshit. To see how a component attitude of an ethical virtue can be fully universal even though it refers to aspects of the environment, we need only to focus on one of the many attitudes that compose honesty. What should our attitude be towards the difference between lying and misleading?

Lying is intending to deceive by making a false assertion. Misleading, by contrast, is intending to deceive by making an assertion that carries a false conversational implicature. For example, if I ask you if you know who broke the window, you might reply "it was already broken when I arrived", even though you do know who broke it. Conversational implicature rests on the assumption that speakers are cooperative. It would have been more cooperative for you to answer that you do know who broke the window. The assumption that you are cooperating therefore grounds the hearer's conclusion that you do not know who broke the window. The hearer ends up with the same false belief as if you had simply lied by asserting that you do not know who broke window. The virtue of honesty must include an attitude towards this distinction between two forms of deception. Does the honest person consider them equally bad, consider lying to be worse than misleading, or consider misleading to be worse than lying?

As members of an epistemic community, we need one another to be credible sources of information. We need to guard against the loss of credibility that occurs if we are caught deceiving one another. Therefore, we should have a strong attitude against lying. This will mitigate against our own lying and will cause us to respond negatively when other people are caught lying. The virtue of honesty therefore includes a strong negative attitude towards lying. For the same reasons, honesty should include a strong negative attitude towards misleading. However, this should frame misleading as less bad than lying. For although being caught misleading does damage a person's credibility, being caught lying damages their credibility more comprehensively. Being caught misleading damages only the credibility of one's conversational implicatures. But being caught lying damages the credibility of one's assertions, thereby undermining the assumption of cooperation, thereby damaging the credibility of one's conversational implicatures, too. This is why we can respond to a misleading answer by pressing a simple question. "Sure, but do you know who broke the window?" The misleader must still decide whether to lie.

It is best if members of an epistemic community do not damage their credibility at all. But if they are going to do so, then it is better that they damage only one aspect of their credibility than damage the whole of it. Given that people will sometimes be faced with strong incentives to deceive, an epistemic community needs a safety valve to relieve this pressure in the least damaging way. We therefore ought to have strong attitudes against both lying and misleading, but which indicate that lying is the worse of the two. These are the attitudes required for preserving our own and one another's credibility. They are attitudes that a member of an epistemic community ought to have, components of the ethical virtue of honesty.[6] These are among a wide range of attitudes that the virtuous person would have, so even the virtuous person would sometimes allow that it is right to deceive, even that it is right to lie rather than mislead. But they will nevertheless have strong attitudes against deception, which allow that misleading is less bad than lying.

If we zoom in to the highest resolution, the beliefs and desires that compose these attitudes refer only to our membership in an epistemic community, the need to preserve credibility, the distinction between assertion and conversational implicature, the possibility of deception, and the inevitability of the temptation to deceive. These are all essential features of the human condition. If treating lying as worse than misleading is indeed the best or only way to manage this aspect of our lives, then this attitude is a fully universal component of honesty even though it refers to aspects of the environment. Unlike ethical integrity, however, honesty might yet involve other attitudes that refer to contingent features of the environment. If it does, then honesty is not a fully universal virtue. Either way, these examples of honesty and integrity demonstrate the two ways in which an ethical virtue, or a component of an ethical virtue, can be fully universal: by not referring to aspects of the physical and social environment at all, or by referring only to aspects of the physical and social environment that are essential to the human condition.[7]

Notes

1 This chapter's summary of attitude psychology is based on four literature reviews by leading attitude psychologists: Cooper (2007); Fazio & Olson (2007); Maio, Haddock, Manstead, and Spears (2010); Maio, Haddock, and Verplanken (2018).

2 Treating the measures of strength in aggregate has the further advantage that it allows us to model this single dimension of an attitude in terms of the strengths of

association between each of its component mental states, as we will see in the following section. If we model attitude strength in this way, then we can explain why these four measures all detect the same dimension of the attitude and why their aggregate predicts the attitude's persistence, resistance, and influence on cognition and behaviour. For the details of this modelling and explanation, see my forthcoming paper "Attitudes in the Personality System: Two Kinds of Disposition" (Webber, in preparation).

3 I have simplified here. Tanesini allows that an attitude can comprise an object, a set of valences, and associations between that object and those valences (in prep, §§ 3.1.1, 3.1.3). This complication is intended to accommodate ambivalent attitudes. On my preferred model, by contrast, ambivalence is explained more simply: the attitude is identical with its base, which includes both positive and negative beliefs and desires about the attitude object. My analysis in this chapter excludes the phenomenon of ambivalent attitudes for the sake of clarity. Nothing hinges on this omission.

4 Moreover, this model makes two predictions that are in principle testable. One is that the consistency of an attitude's expression across different kinds of situation is a matter of degree that correlates with the overall strength of the connections between the items in its base. The other is that the situations in which a generally strong attitude is not elicited even though it would seem appropriate are ones in which the aspects of the attitude's base relevant to the situation are only weakly associated with the rest of the base. These predictions are not made by the model of an attitude as a distinct mental state caused by its base.

5 The virtue of ethical integrity requires this to be an unrestricted generality. But the related virtue of professional integrity seems to be fully universal, too, even though its component attitudes have a restricted generality of application. That is, professional integrity comprises constancy and fidelity with respect to one's professional commitments, a strong commitment to getting things right in one's professional life, and a strong attitude towards balancing that constancy and fidelity with that drive to get things right. If this is to be genuine professional integrity, all of its component attitudes must range over all of one's professional commitments, whatever those might be. The professional integrity of a surgeon is the same as that of a teacher, therefore, at every degree of resolution, even though their professional environments are very different and their specific professional commitments have very different attitude objects.

6 This section summarises my more detailed argument that we ought to consider lying to be worse than misleading (Webber, 2013, §§ 1–3). My reasoning has been misread as an act-consequentialist argument either that lying is worse because it removes the option of asking further questions (Rees, 2014, pp. 62–63; Cohen, 2018, pp. 69–70) or that lying is worse because it causes greater harm to the linguistic community (Berstler, 2019, n. 28; see also Pepp, 2019, n. 44 and § 5). However, the argument's conclusion is not that lying is worse than misleading. It is that we ought to have a stronger negative attitude towards lying than towards misleading (Webber, 2013, pp. 654, 658). Of course, if we do have that attitude, we will say that lying is worse than misleading. But the reason we should say that lying is worse than misleading, according to this argument, is that the virtuous person, who has the attitudes we ought to have, considers lying to be worse than misleading. This is virtue ethics, not act-consequentialism.

7 This chapter derives from my keynote address at "Virtues: Local or Universal?", the seventh annual conference of the Jubilee Centre for Character and Virtues, at Oriel College Oxford in January 2019. I am very grateful to the organisers and participants of that conference for the opportunity to develop these ideas through that discussion. I am also grateful to Catherine Darnell, Kristján Kristjánsson, and Alessandra Tanesini for helpful comments on the first draft of this chapter.

References

Axsom, D., & Cooper, J. (1985). Cognitive dissonance and psychotherapy: The role of effort justification in inducing weight loss. *Journal of Experimental Social Psychology*, *21*, 149–160.

Berstler, S. (2019). What's the good of language? On the moral distinction between lying and misleading. *Ethics*, *130*, 5–31.

Cohen, S. (2018). The moral gradation of media of deception. *Theoria*, *84*, 60–82.

Cooper, J. (2007). *Cognitive dissonance: Fifty years of a classic theory*. Los Angeles, CA: Sage Publications.

Fazio, R. H., & Olson, M. A. (2007). Attitudes: Foundations, functions, and consequences. In M. A. Hogg & J. Cooper (Eds.), *The Sage handbook of social psychology: Student edition* (pp. 139–160). London: Sage Publications.

Holland, R. W., Verplanken, B., & van Knippenberg, A. (2002). On the nature of attitude-behavior relations: The strong guide, the weak follow. *European Journal of Social Psychology*, *32*, 869–876.

Maio, G., Haddock, G., Manstead, A., & Spears, R. (2010). Attitudes and intergroup relations. In J. Dovidio, M. Hewstone, P. Glick, & V. Esses (Eds.), *The Sage handbook of prejudice, stereotyping and discrimination* (pp. 261–275). London: Sage Publications.

Maio, G., Haddock, G., & Verplanken, B. (2018). *The psychology of attitudes and attitude change* (3rd ed.). London: Sage Publications.

McMullin, I. (2018). *Existential flourishing: A phenomenology of the virtues*. Cambridge: Cambridge University Press.

Nussbaum, M. C. (1993). Non-relative virtues: An Aristotelian approach. In M. C. Nussbaum & A. Sen (Eds.), *The quality of life* (pp. 242–269). Oxford: Oxford University Press.

Pepp, J. (2019). Assertion, lying, and untruthfully implicating. In S. Goldberg (Ed.), *The Oxford handbook of assertion*. Oxford: Oxford University Press. https://doi.org/10.1093/oxfordhb/9780190675233.013.40

Rees, C. F. (2014). Better lie! *Analysis*, *74*(1), 59–64.

Rees, C. F. (2016). A virtue ethics response to implicit bias. In M. Brownstein & J. Saul (Eds.), *Implicit bias and philosophy, volume 2: Moral responsibility, structural injustice, and ethics* (pp. 191–214). Oxford: Oxford University Press.

Rees, C. F., & Webber, J. (2014a). Automaticity in virtuous action. In N. Snow & F. Trivigno (Eds.), *The philosophy and psychology of character and happiness* (pp. 75–90). New York: Routledge.

Rees, C. F., & Webber, J. (2014b). Constancy, fidelity, and integrity. In S. van Hooft (Ed.), *The handbook of virtue ethics* (pp. 399–408). Durham: Acumen.

Tanesini, A. (in preparation). In *The mismeasure of the self: A study in vice epistemology*. Oxford: Oxford University Press.

Webber, J. (2013). Liar! *Analysis*, *73*(4), 651–659.

Webber, J. (2016). Instilling virtue. In A. Masala & J. Webber (Eds.), *From personality to virtue: Essays on the philosophy of character* (pp. 134–154). Oxford: Oxford University Press.

Webber, J. (in preparation). Attitudes in the personality system: Two kinds of disposition.

5 The essential sociality of Aristotelian virtue

Jennifer Rothschild

Introduction

In this chapter, I argue that Aristotelian practical virtue is essentially social. By "essentially social", I mean more than just the thought that social goods play a critical role in the development and maintenance of individual virtue. I mean that virtuous practical activity just is a marker of properly navigated sociality. Generosity is appropriate action in the domain of giving, which just is giving that appropriately situates my good with the good of others. I think we can follow this kind of understanding about generosity out to similar thinking about even virtues such as moderation. Completed virtue may be something we hope to see in individuals, but its being and goodness are structured by the social nature of excellent human activity.

The plan of the chapter is as follows. In the first section, I address the sense of "essentially social" at work in the paper. In the second section, I argue that Aristotle explicitly endorses an essentially social understanding of virtuous practical activity. In the third and fourth sections, I turn to particular virtues of character, arguing that the Aristotelian virtue ethicist has good reason to make sense of these virtues as essentially social. The third section focuses on virtues like generosity, virtues which wear their sociality openly. The fourth section takes on what I consider to be the hardest case for my account: moderation.

From the point of view of the question motivating the present volume, about whether virtues are local or universal, the argument in this chapter makes the following contribution: there exists a core set of human virtues which are fully universal. The universal virtues are those which are, in an Aristotelian framework, tied to our nature as human beings and unified in the best person; here I add to the thoughts about nature and unity the claim that the universal virtues are essentially social.

In what sense "essentially social"?

When I say that the Aristotelian practical virtues are best understood as essentially social, I intend "essentially social" in a strong sense: in my view, virtuous activity is, in kind, a social achievement aimed at a common good. Individual achievements in virtue, while not unimportant, can only be made sense of in the light of this aim.

Essential sociality of the relevant kind outstrips a weaker thought about sociality, one in which the social is critical for the development and maintenance of individual virtue. The weaker thought can highlight the importance of families, friends, school, and political arrangements for the production of virtue, noting the ways in which they provide irreplaceable opportunities for learning and doing virtuous activity (which may, in turn, have good consequences for the shaping of the relevant social). On this understanding, however, the essential sociality is limited: individual virtue is dependent upon the social merely developmentally. In other words, the social is prior to the individual achievement, in the sense that practices support the achievement's coming into being, but no further dependency need hold.

Consider the following case as an illustration of the weaker relation of developmental sociality: suppose a speaker at a scholarly conference receives an award for doing good philosophical work. The award is an achievement that depends on a social world as frame for the opportunity. People and practices like teachers, commentators, conference organisers, and so on, helped support an individual on the way to achieving that good. The achievement may have consequences for the relevant social, too, as when a paper influences the ideas of her fellow scholars.

Now suppose, by contrast, that our scholar was recognised for collegiality at a conference. Understanding what happened in this case – understanding what our scholar did when she acted collegially – runs quite differently than the case just under discussion. In the case of collegiality, what our scholar did is do her part in a conference well in a way that facilitated the conference itself, which is a project of many people. In kind, the achievement is one that does not make sense without connecting that individual in the right ways to her shared activity with other people, connecting her relevant good to their relevant good. What our scholar did in this case is occupy a place in a shared life and a shared good which, when understood aright, can in turn highlight what it is she did on her own that contributed to it. In other words, understanding the shared life here has consequences for how we understand the individual activity and achievement, not the other way around.

In the case of collegiality, the social is prior to the individual not developmentally but conceptually, and hence the dependency of the individual achievement on the good of sociality is a dependency of intelligibility.[1] My argument will be that virtue is essentially social in this second sense. As in the case of collegiality, we cannot understand the kind of achievement virtue is without reaching for an understanding illuminated by shared, good activity with others.

Aristotle's *Nicomachean Ethics* on the essential sociality of practical virtue

The argument that virtue is best understood as essentially social is intended to be an Aristotelian argument. In the unfolding of the argument, my overall aim will be to do virtue ethics that is faithful to Aristotle on the main theoretical points, though I will operate at times nearer and at times farther from the text of Aristotle, and I will even, in the fourth section, note a point on which I think we would do best to depart from Aristotle. Readers who are interested in the overall aim, but not in Aristotle himself, will not lose out by skipping to the third section.

In the current section, I aim to show that Aristotle endorsed the claim that practical virtue is essentially social in the way I mean it. To do this, I focus on the *Nicomachean Ethics* rather than the *Politics*, for two reasons:[2] first, because the *NE* is the anchoring text of virtue ethics, and is therefore the text from which the usual emphasis on individual virtues is derived. Second, I work from the *NE* because it is conducive to consideration of the small-scale social – an individual's interaction with family, friends, and neighbours – as opposed to the large-scale social of politics. In my view, the small-scale social is the most immediately productive for thinking about why what looks individual may, in fact, be social.

Aristotle argues in *NE* I.7 that our good is a good belonging to our kind. Human beings are souled creatures, and the good of a souled creature consists in its activity. Human souls are also rational, and so they are best when they are in a condition of activity that is organised by reason. Aristotle says it this way:

> [I]f all this is so, and a human being's function we posit as being a kind of life, and this life as being activity of soul and actions accompanied by reason, and it belongs to a good man to perform these well and finely, and each thing is completed well when it possesses its proper excellence: if all this is so, the human good turns out to be activity of soul in accordance with excellence (and if there are more excellences than one, in accordance with the best and the most complete).
>
> (Aristotle, Rowe, & Broadie, 2002, p. 102)

This passage argues that we should understand a good human being as one whose soul is characterised by the right sort of activity (rational activity) and who is engaged in the right sorts of actions (actions accompanied by reason). This is the person we can call excellent, or virtuous. The human being in such a condition of excellence is someone who is doubly marked as completed: she is completed as the kind of thing that she is – as a human being. She is also characterised by her completed virtues (virtues in their best form), the most important of which will be the virtue that is, by nature, itself most complete.[3]

The passage suggests that if we want to understand human virtue in Aristotle, we will do well to look to virtue at its best and define it by its completed form. For the purposes of my argument, it is *practical* virtue[4] that is immediately relevant, as practical virtue is the stuff of ethics and the kind of virtue I claim is essentially social. So, we should focus the suggestion this way: if we want to understand practical virtue in Aristotle, we will do well to look to practical virtue at its most complete. For that is the form of the virtue that is the standard of practical achievement for individuals of our kind, and the standard is that which makes intelligible the practical activity of actual, individual humans.

There are two good ways into Aristotle's understanding of complete practical virtue in the *NE*, reflecting the doubly marked completion mentioned previously. One is to concentrate on the practical completion of the human soul. What, according to Aristotle, is the best condition for the practical soul of a human being? Aristotle says in *NE* VI that excellent activity of this kind belongs jointly to the two practical parts: the deliberative (rational) part of the soul, whose excellence is practical wisdom, and the desiring (non-rational) part of the soul, whose excellences are the character virtues like justice, generosity, courage, moderation, and the rest. He is clear that practical virtue depends on the achievement of these virtues as a unified set (Aristotle et al., 2002, pp. 110, 188–189).[5] Focusing on the virtuous soul, then, will take us through the argument for the unity of the virtues. I do think – and I hope this will become clear in the third and fourth sections of the chapter – that the unity thesis and the essential sociality thesis hang together, and that understanding either aright will depend on understanding the other.

A second way in to a thought about completed practical virtue is to focus not upon the soul but upon the kind of virtue that practical virtue is, as well as what it would take to complete a virtue of that kind. In other words, what is it for practical virtue in particular, as opposed to some other virtue, to be at its most complete? Aristotle argues in Book X that by contrast with contemplative activity, practical activity of soul is second best, and its completion depends upon pursuit of an end it does not contain (Aristotle et al., 2002).[6] This kind of activity is not self-sufficient and divine, but dependent and human;[7] continuous pursuit of it comes through the perfection of the activity of living in the world with other human beings. While contemplative activity is the sort of thing we can do when we are alone, practical activity needs others: "the just person will need people to be objects of, and partners in, his just actions, and similarly with the moderate, the courageous, and each of the other types" (Aristotle et al., 2002, p. 251).[8] This is a stronger claim than the thought that virtuous actions come more easily when acting with other people, for Aristotle says in this same stretch of text that contemplative activity, too, benefits from doing it with other people. The important distinction is that while it is better to contemplate together, it is possible to do so on one's own, and what is more, it is possible to do so excellently on one's own. Practical virtue does not share in this possibility. He repeats the point in this form:

> The person engaged in reflection needs none of these sorts of things, at least for his activity . . . but in so far as he is a human being, and shares his life with others, he chooses to do the deeds that accord with excellence, and so he will need such things for the purposes of living a human life.
>
> (Aristotle et al., 2002, pp. 253)[9]

Aristotle emphasises here that the kinds of actions that make us good human beings mean a shared life with others, and so being virtuous as "one of us" is living a life of that kind. Practical activity in particular is the kind of thing whose excellence depends on the shared life. Here we see the twin pillars of Aristotelian naturalism – that our flourishing is constituted by a certain kind of activity of soul, and that we are, by nature, social animals – joined together as one, linked by the social character of practical activity of soul.[10]

The essential sociality of generosity and its familiars

Suppose we grant that Aristotle thinks virtue is essentially social. We should wonder next whether this part of his view is worth taking up, and what it would mean for our reading of virtue if we do. In this section, I begin a two-part argument, which aims to show that the individual character virtues make a great deal more sense if we think of them as socially constituted excellences aimed at a shared good. I concentrate on the virtues that are most readily understood as social, and in the following section, I turn to character virtues that are less obviously dependent on sociality for intelligibility.

Consider, for example, the virtue of generosity, and in particular, consider cases of failing at generosity by giving too much.[11] Aristotle thinks that giving the right amount, in the right way, at the right time, and so on, is one kind of getting it right between human beings. Missing a virtue by going over the mean (giving too much) is to act in a way that is itself a failure to act according to that virtue (a failure of generosity) (Aristotle et al., 2002, pp. 112–113, 115).[12] Readers of Aristotle sometimes have intuitive trouble seeing giving too much as a failure of generosity, but this trouble disappears if we pick out cases of giving to excess in which it is clear that the good of the giver or others is jeopardised by the excess. In other words, the category of acting generously comes visible, and works in the way Aristotle suggests it should, when we highlight the essentially social nature of generosity. Getting generosity right depends upon activity aimed at a shared good; failing one means failing the other.

Consider a case in which someone gives to a point beyond her own good, the excess draining her resources to protect and care for herself. Perhaps she is a vulnerable person who is being exploited, giving excessively to someone taking advantage of that vulnerability. This kind of giving is bad for the giver. It is working against her living well, and we feel compelled to say that this cannot be what it is for her to be generous. This is a "too much giving" that can only be a failure of the very virtue it means to enact. Similarly, if someone gives too much in a way that works against the good of others, we can see this, too, as a failure of generosity. Suppose, for example, that a person has a favourite grandchild, and showers that child with gifts on every possible occasion, while a less favoured grandchild looks on, receiving nothing. This grandparent is not generous; rather, this is a case of giving too much in a way that fails to be generous. (It is a case of giving too little, too, to the other child in question.)

We can group the cases of the exploited giver and the grandparent together, using them to mark an understanding of generosity that is properly Aristotelian in spirit. When I give too much in the way relevant to generosity, I have used my action to move against my own good or the good of others. Giving generously means neither giving too much at my own expense nor holding too much for myself at the expense of someone else.[13] Notice here that there is nothing that it is to be generous that can be abstracted away from navigating human beings and their good. The getting it right of giving just is a kind of handling well of the good of the humans involved. When giving goes well, it binds us together and promotes flourishing in the space of a life lived with others. When it does not, it harms us or others and alienates us from one another. There is no sensible way to understand giving as a virtue – as generosity – that does not situate it as a matter of getting it right between people, or if you like, getting it right for me in a life whose good is shared with others.

As we know from Aristotle, getting it right is not only a matter of doing neither too much nor to little, but it is also a matter of acting in the right manner, toward the right targets, at the right time, and so on (Aristotle et al., 2002, p. 117).[14] In the previous examples of generosity, some of the other evaluative markers are already visible. The person giving to someone who exploits her is giving to the wrong target. The one who wants to give something extra to one child is not doing so at the right time (while the other child is present) or in the right manner (showering the one child with gifts and ignoring the other). These failures of manner, target, and so on are failures of the relevant kind of sociality, too. They are failures to act in a way that properly navigates my good and the good of others – failures of the giver to, through her action, take up the right place in relation to her recipient.

We can, I argue, give similar accounts of the essential sociality of other important character virtues. There is no way of making sense of justice without relying on an understanding of what it would be for someone (either me or someone else) to have a share in something in which some other(s) also have a share. There is no way of making sense of honesty that does not rest fundamentally on what it would be to get truth-telling to others right. There is no way of making sense of kindness without casting its fulfillment as human beings treating one another appropriately gently.[15] And I think there is no way of picking out why failures of these virtues are virtue-failures without running that understanding thorough their essential sociality, as being about whether I am living well as a human being appropriately attuned to other human beings.

For virtues like courage, or especially moderation, the sociality is less transparent. I will save these for the next section, where I concentrate on moderation in particular. Before we move on to harder cases, though, I want to point out two further features of the cases of generosity we were just considering. We will need these features in order to see the range of what essential sociality of virtue gets us onto, we will need them in order to read the harder cases well, and finally, I argue, we have reasons for thinking both features mark something worth noticing in Aristotelian virtue ethics beyond the way the features figure into arguments about sociality.

One feature of the previous cases worth highlighting is the ways in which generosity runs into concerns of the other virtues. There is clearly something unjust and unkind about the way the first recipient takes advantage of the vulnerable giver. We might imagine that in order to do better going forward, our vulnerable giver will need a recalibration of her courage, or will need the kind of patience that does not simply wait but knows when enough is enough. If our giver is going to do better, her improved position will necessarily involve activity that is more than simply more generous. Getting it right will include doing better with other virtues, as well. This feature picks up on Aristotle's argument that the character virtues are unified. Minimally, this means that they are a set that are achieved and pursued together (Aristotle et al., 2002, p. 189).[16] As I said in the second section, I think the unity thesis and the essential sociality thesis travel together. There is much to say about this claim that goes beyond the argument here, but I think it counts as a strength of the argument of this chapter that it can accommodate thinking in terms of multiple character virtues at once, seeing how they might reinforce one another. When we see the essential sociality of generosity, we see some of the ways its concerns point to the concerns of other virtues. This is not surprising, given that we are understanding the other character virtues as similarly social in activity.

Second, the previous examples of generosity lend themselves to a fairly expansive reading of generosity, and I would argue that expansive readings of individual virtues are a good thing for doing virtue ethics. By "expansive reading", I mean understandings of individual virtues that allow them broad reach, and which make them widely relevant beyond the narrow range of their most obvious, paradigmatic cases. Perhaps the obvious paradigmatic cases of generosity are those involving giving money or gifts. But we can easily imagine the grandparent in the earlier example giving her time and attention to a child, likely in this case more of it to one child than the other. We would expect the exploited giver to be more willing to give her recipient the benefit of the doubt on matters she should not. The giving of time, attention, and listening are kinds of giving to one another that can and should be evaluated in terms of generosity. They are actions whose ways of going right and wrong are illuminated by thinking of them in virtue terms. Essential sociality of virtue is a way of framing individual virtues that is compatible with, and more, possibly demanding of, expansive readings of individual virtues as opposed to narrow ones.

The essential sociality of moderation (and its familiars)

In this final section of the chapter, I turn to moderation, the character virtue whose sociality is arguably the most difficult to see. Consider the person with a healthy and properly habituated set of appetites for food. Why should we see this condition as dependent on other people's good in the sense relevant for the argument at hand? The weaker sense of sociality is easy enough to see here: living a shared life with others influences the individual's opportunities to develop moderation, and the individual's development of moderation has consequences for the functioning of the social to which she contributes. But the stronger claim – the claim that we cannot even make sense of moderation except through the achievement of a good that is shared – seems a difficult one to make.[17]

Moderation is a virtue that is interestingly different from the virtues in the previous section; virtues like generosity, justice, honesty, and kindness. I want to discuss two such differences,[18] both of which are conducive to the thought that moderation is not social like the other virtues.[19]

One important difference is that we can get a bit farther with understanding moderation outside the social than we can with some virtues. When we ask what it is to act generously, for example, we cannot even get the question off the ground without reaching for human sociality. To recall the example from the first section, generosity is like collegiality. If we want to know what collegiality is, there is nowhere to go unless we reach for other people immediately. The same is not true for moderation: we *can* ask what it would mean to have our desires for food in order without having to immediately reach for other people. We may even make some progress in figuring out partial answers to the question. For example, how much food leaves one hungry or makes one sick is an experiment we can do alone.

But this line of thought, if it goes anywhere, does not take us far. We are sophisticated, rational, desiring beings; we are beings in whom appetitive desires are far from exhausted by the biological needs of the body. As we follow our desires out from those of basic need to those more sophisticated, it becomes increasingly pressing that we integrate our appetitive desires with our other desires. We are not acting moderately if we put ourselves in the position of needing to skip a day's meals to finish a paper, of if we meet mom's cooking over the holidays with an appetite that requires two naps a day. Neither strikes us as the actions of someone who is appetitively ordered, and both are ways of acting that divide us from others and deprive us and relevant others of our good, shared life. Being truly moderate in a way that is not just general (don't eat to sickness) and trivial (ignoring the complexity of the human kind, which cannot be the stuff of virtue) requires an ability to navigate my good, and ours, in all its complexity, promoting and not disrupting that life.

As we saw in the third section with generosity, concerns of moderation quickly run into the concerns of other virtues: it is not good to eat in such a way to keep me barely on the safe side of sickness if this contributes to my eating more than my share. Such a way of being would interfere with my ability to act justly. I would not be moderate if I chose to spend my days on my couch eating pizza with a kind of failing courage to face the world, even if I roughly ate an appropriate amount of pizza. In such cases, I fail to be moderate because I pursue food in ways that isolate me from others in refusal of sociality. Of course, eating to almost sickness or living a life of couch pizza are also actions that are not good for me individually: they are not good for my body, or for the stability of my desires. But what happens to me in consequence of my failed virtue activity does not explain the failed activity itself. The mistake in virtue that I make when I act this way is acting against the good of a shared life which itself marks my good. In doing this, I act immoderately. Here we see the force of the points about unity and sociality joined together: to truly understand what moderate activity is, we have to be able to let it meet the other virtues and be refined in the light of them; and there is no way to work from my basic desires to a full story about what is good for me without anchoring that story in a good I share with others.

A second important difference between moderation and virtues like generosity is that moderation's domain of action is often defined by the limited number of objects its desires are said to take, most notably the objects of food, drink, and sex.[20] This restriction we owe to Aristotle directly (Aristotle et al., 2002, p. 138; see also Broadie, 2002, pp. 54–57).[21] Fixing our attention exclusively on three objects of bodily desire makes it easy to think we should treat moderation as an individual matter as opposed to a social one.

We owe moderation a more expansive casting than Aristotle allows. To read moderation expansively, I argue, we need to make a shift to reading moderation as the perfected activity of the appetite, including, but not limited to, our appetitive desires for food, drink, and sex. A full accounting of this shift is beyond the scope of this paper, but roughly what is needed is to do what we do with the other virtues: we define the virtue not by the characteristic object of the virtue but instead by a rich accounting of the domain and manner of that virtue's activity. So instead of treating the appetites as desires that come to us by way of certain objects pursued as pleasant, we instead concentrate on the appetitive activity. When we do this, we start to see that appetitive desires have a particular character:[22] they arise from perceived necessity and seek pleasure in satisfaction. Close on the heels of any satisfaction of them are desires born anew; this feature makes them particularly likely to get out of control, to seek new outlets for satisfaction, and to encourage their bearer to overlook better desires and pursue to excess. On the shift I am endorsing, moderation is the virtue that is achieved when we have ordered this kind of desiring (appetitive desiring), i.e., when we are in the best condition with respect to that kind of desiring activity. The shift allows us to unite the three usual objects of moderation, offering the beginnings of an account of both why the three belong together as a group as well as why our way of dealing with that group is pressing for virtue. The shift also opens the way for us to include other appetites as relevant for moderation. If we can find candidate objects that inspire the same kind of activity as food, drink, and sex, we have reason to expand the virtue to include these new objects.

An appetite for money is a good candidate for inclusion under moderation in my expanded sense.[23] Notice that the appetitive activity of the pursuit of money has much in common with our desires for food, drink, and sex. The styles of pursuit have a similar grasping quality. When they overreach, they lose track of pursuing the object as it is needed or useful. Excessive desire for money is greedy, which strikes us as quite similar in personality to insatiable hunger, or thirst, or lust. A human being's immoderate appetites for money can interfere with her flourishing in ways that divide her from others, just as can immoderate desires to, say, drink alcohol. There is a similarity to the ways in which we act badly when excessive appetites for money or for sex overtake our better judgment. In short, I think we have good reason to understand the domain of moderation as including all the appetites we might have a hard time domesticating, all the appetites that threaten to grow bigger than they should. Seen in this light, desires for money look to belong in this domain right alongside the other three.

There is good precedent for including money under moderation. In *Republic*, for example, Plato defines necessary appetites in such a way as to include both the things we need for survival and also "those whose satisfaction benefits us" by contributing to our well-being. The latter extend beyond basic bodily needs to other needs of flourishing, such as the moderate pursuit of delicacies and, he explicitly notes, money (Plato, Grube, & Reeve, 1992, p. 229).[24] Plato makes a particularly compelling case for the ways in which human beings are endangered by their desires to accumulate money, too (Plato et al., 1992, p. 118),[25] and he clearly allows a vicious pursuit of money to count as immoderation (Plato et al., 1992, pp. 118, 223–224).[26]

Expanding the domain of moderation in this way, beyond the body to complex social objects we can lust after, makes good sense of the psychology of human beings and of a particular kind of threat to our flourishing. The expanded domain makes it even harder, moreover, to hold onto the thought that moderation is a matter understandable outside sociality. Nothing is more immoderate nor more disruptive of the common good as appetite run amok. It is hard to imagine how we would fully capture this truth by thinking of moderation as a concern about individuals that then bumps into other people, as opposed to thinking that we could only pick out individual moderate activity and its way of going right when we structure its good in terms of social aims.[27]

In the end, it may be that some differences between moderation and virtues like generosity persist. Perhaps we should say that moderation is not quite like collegiality. Rather, the relevant comparison might be a team activity with role players, like understanding being a good point guard in basketball. But if this is right, the difference is not a great one. For maybe we can say *something* about the skills of a good point guard without talking about the whole team and even the whole game (likely this will be something trivial and general about good ball handling, quick feet, and so on), but we cannot say much, and we certainly cannot understand what it is to be a good point guard, without understanding the role the good point guard plays in pursuit of a team basketball victory.[28] And though it may be true that once we know what a good point guard is, we can say an individual point guard can practice aspects of her game on her own, this does not change the fact that what she practises on her own is excellent basketball playing only in a limited and derivative sense.[29] The differences between the analogies of collegiality and basketball are not uninteresting, but neither are they a reason to deny that properly understood, we cannot so much as make sense of perfected, moderate activity without reading it as one domain of action in a well-navigated life with other people, one which situates the importance of their good alongside my own. If I am moderate, I have regulated my own appetites as part of taking up my proper place in a shared life with other human beings. Thus, moderation, like generosity and the others, is essentially social.

In closing, let me take a moment to connect this chapter to the theme of the volume concerning the question of the locality or universality of virtue. The argument of this paper is an attempt to show the essential sociality of those character virtues which are, in an Aristotelian framework, necessary and jointly sufficient for being a fully virtuous person. In other words, they are: (1) virtues marking the good human being necessarily, by nature; and also (2) virtues which are a necessary part of the goal of virtue unity. This excludes some "virtues" Aristotle mentions, such as magnanimity; though essentially social in shape, magnanimity is not a virtue anchored to the goodness of a human being as such, nor is it required by virtue unity. The virtues in question here are the core virtues of character, virtues such as justice, generosity, moderation, and courage, as well as some that Aristotle did not name, such as honesty and patience and kindness. These core virtues, those which I aim to show are essentially social, are connected to what we are and what makes us good in ways that make them fully universal: relevant to all of us, worth pursuing for each person in pursuit of the good, and in particular, for the good life enjoyed jointly with others. Such an argument leaves plenty of room, of course, for local ways of navigating and instantiating these core virtues, as well as for the introduction of more locally relevant secondary virtues.

Notes

1 In the case of virtue, the dependency is importantly metaphysical as well.
2 In the *Politics*, it is still clearer that completion of the individual depends upon the priority of the good of the whole.
3 Human virtue is the sort of thing whose intelligibility attaches in general to our kind and only secondarily to individuals who are examples of that kind.
4 Practical virtue includes character excellence and practical wisdom, as outlined following.
5 *NE* I.13 and VI.12–13.
6 *NE* X.8 1178a9 and *NE* X.7 1177b3. Contemplative activity is our highest achievement; it contains its own end, enabling continuous and self-sufficient pursuit of completion (*NE* X.7 1177a18, b1, and b20–25).
7 Contemplative activity is likened to something divine (*NE* X.7 1177b25–32 and *NE* X.8 1178b22ff). Practical activity is distinctively human activity. (*NE* X.8 1178a9ff). Practical activity's "humanness" here is picking out the nature of practical activity as shared with others.
8 *NE* X.7 1177a31–33.
9 *NE* X.8 1178b3–7.
10 A more complete version of this argument would connect with Aristotle's arguments about friendship.
11 Often, I think, the centre of a virtue's goodness comes into view by way of getting clear on what is going wrong in a failed case. Cases of excess are particularly fruitful for this.
12 *NE* II.2 1104a20–27 and II.6 1106b19–23.
13 Holding too much for myself includes holding too much for one I take to be an extension of me, such as my favourite grandchild.
14 *NE* II.6 1106b21–23
15 Perhaps we can direct something like generosity or kindness or honesty at ourselves. If it makes sense to say I can "be honest with myself", however, that possibility will be derivative on the sociality of honesty, for it depends upon my treating myself as both agent and other. I do not think there is any sensible way to cast treating oneself justly.
16 *NE* VI.13 1144b35–a1.
17 When it comes to the virtue of moderation, Aristotle is not entirely clear or consistent about whether he endorses the weaker or stronger sense of sociality. I attribute this to his misunderstanding of the scope of moderation, which I address later in this section.
18 Another difference I discuss shortly is that virtues like generosity are arguably only sensibly executed toward other human beings, at least in the primary sense. Moderate activity, on the other hand, does not need to be aimed at separate a human target. I raise this difference only to suggest we ought not be distracted by it. What we need to know is not whether a person can act moderately while alone, but whether the goodness of one's moderate activity is visible to us (including while she is alone), independently of a primary understanding of moderation which is social in nature.
19 Courage has a great deal in common with moderation, and it would be interesting to consider the two virtues together here. I omit discussion of courage in the interest of space.
20 It is almost always unclear how broad Aristotle intends a virtue to be. Generosity has to do with wealth or anything that can be bought with it, which leaves room for an expansive reading (*NE* IV.1 1119b25–26). He gives us a varied and rich sense of the phenomena associated with courage: from war-making to pain to other sorts of external threats, human beings will need courage to face their perceived enemies well (*NE* III.8).

21 *NE* III.10 1118a1.
22 On my reading of Plato, this casting of appetite is more his than Aristotle's.
23 As Broadie notes, Aristotle acknowledges an appetite for money as an unnecessary appetite to which moderation need not apply (Broadie, 2002, pp. 54–57). Here I suggest instead that we expand the domain of moderation to include appetites such as the appetite for money. I think we can expand Aristotle in this way without violating anything central in his thought, and in fact, I think we do better by him to elevate moderation alongside the other character virtues.
24 *Republic* 559a–c. See also the closing lines of *Phaedrus* at 279c, where Socrates offers this prayer: "As for gold, let me have as much as a moderate man could bear and carry with him" (Plato, Nehamas, & Woodruff, 1995, p. 86).
25 *Republic* 442a.
26 For Plato, a human is moderate when the parts of the soul agree that reason should rule (442c–d). The oligarch violates that order – in letting his appetites for money overtake his better nature, he is twisted into an immoderate creature. (553c–d).
27 We might again take our cue from Plato: in its full development, moderation is social to a rather extreme degree, more like the virtue of justice; where as some virtues (wisdom, courage) are virtues of particular parts of the soul, moderation and justice are first and foremost qualities of the interactions of the whole soul (and city).
28 This suggests moderation will come most visible when we consider its place on the team: when we unify it with character virtues and practical wisdom, and also when we grant it its place in the political.
29 A parallel case here is Robinson Crusoe, who continues to carry on certain aspects of moderation even once he is living all alone. See fn 18.

References

Aristotle, Rowe, C. J., & Broadie, S. (2002). *Aristotle: Nicomachean ethics*. Oxford: Oxford University Press.

Broadie, S. (2002). Philosophical introduction. In C. J. Rowe (Ed.), *Aristotle: Nicomachean ethics* (pp. 9–91). Oxford: Oxford University Press.

Plato, Grube, G. M. A., & Reeve, C. D. C. (1992). *Republic*. Indianapolis: Hackett Publishing.

Plato, Nehamas, A., & Woodruff, P. (1995). *Phaedrus*. Indianapolis: Hackett Publishing.

6 Neo-Aristotelianism, the universality of virtues, and the challenge of germline genome editing

Nancy E. Snow

Introduction

Neo-Aristotelianism offers a picture according to which cultural variability is the factor most responsible for virtues being local, that is, relativized to specific times and places; and nature, that most responsible for their being universal, that is, shared in common by all human beings. Call this the "traditional picture". On this picture one might say, for example, that the Korean virtue of *nunchi*, which is a kind of social intelligence that is thickly embedded in Korean culture and mores, is local, developed within, pertaining to, and meaningful within Korean society, whereas a virtue such as benevolence is universal, appearing in all cultures, and meaningful across cultures, albeit expressed in differing ways.[1] If I, a Westerner, were to go to Korea, I would not be able to identify expressions of *nunchi*, and would likely violate social expectations and norms connected with that virtue, because it is so deeply embedded in the fabric of Korean culture. By contrast, I would be able to identify and appreciate expressions of benevolence because of its ubiquity and universality across cultures. Neo-Aristotelians can avoid the specter of relativism here by contending that human capacities are the stable features grounding the development and expression of virtues – *nunchi*, though developed in a specific cultural context, depends upon the same human capacities for reason, desire, and emotion as benevolence. Thus, even though virtues might vary locally in cultural expression – some might even be culturally specific – our shared capacities provide a natural basis in which the universality of most virtues can be grounded.

Even the briefest glance into evolutionary perspectives on heritability suggests that virtues are more radically local, that is, rooted in the contingencies of human evolutionary development as it occurs within specific contexts over time, than the traditional picture recognises. Manipulations of nature through technologies and cultural factors have produced changes in human capacities. For example, the advent of language and literacy produced populations that were more tolerant, self-aware, and attuned to art and beauty than their illiterate, more violent predecessors. These, and heritable changes wrought through stable social structures, such as education and religion, arguably produced changes in the kinds of goods and virtues constitutive of human flourishing.[2]

Neo-Aristotelian ethical naturalists have been unwilling to engage with contemporary evolutionary theories, including theories of heritability, and thus have not confronted the complexities that evolutionary heritability raises for their accounts in purporting to explain why and how virtues are, or can be, universal. This evasion, untenable as it is, is made even less so by the advent of germline genome editing – a new kind of technology that can produce heritable changes in members of many kinds of species. The speed, flexibility, and kind of heritability made possible by germline genome editing threaten the neo-Aristotelian stance on the universality of virtues with a new urgency. Aside from issues of metaethics, germline genome editing forces us to consider with care its implications for human goods and virtues at the level of normative ethics.

In the first section following, I sketch the main contours of neo-Aristotelian ethical naturalism and discuss scientifically based problems that arise for it, including a challenge to the universality of virtues. In the next section, I offer a brief overview of developments in genome editing. In the final section, I suggest a way in which ethical naturalists might respond to the metaethical challenge developed in the first section following, and discuss both metaethical and normative questions posed by germline genome editing.

Scientifically based problems for neo-Aristotelian ethical naturalism[3]

Several Aristotelians adopt ethical naturalistic approaches to meta-ethics.[4] I take the Hursthouse/Foot/Thompson conception to be the main one, constituting a more or less unified approach, featuring points of influence, overlap, and coherence. Here I focus on Foot and Thompson.

The central idea of this conception is to show the seamlessness of moral normativity with natural normativity. That is, evaluations of the moral goodness and badness of humans are said to have the same conceptual structure as evaluations of the goodness and badness of plants and animals. Early adumbrations of this approach can be found in work by Anscombe and Geach, and were developed in papers by Foot, in which she argues against moral subjectivism, especially R. M. Hare's version of prescriptivism.[5] Despite this, an early paper of Foot's, "Morality as a System of Hypothetical Imperatives", embraces the notion that morality is essentially grounded in desires or interests. In later work, she repudiates that view, returning to the Anscombe/Geach approach.[6] In Foot (2002a, p. 199), she invokes Anscombe's notion of an "Aristotelian necessity": "that which is necessary because and insofar as good hangs on it". We use this idea, Foot (2002a, p. 199) contends, "when we say that it is necessary for plants to have water, for birds to build nests, for wolves to hunt in packs, and for lionesses to teach their cubs to kill". These judgments provide standards by means of which we can identify individuals that are good of their kind and those that are defective. Since it is the nature of wolves to hunt together in packs and the nature of dancing bees to lead other bees to sources of nectar, we can say that a free-riding wolf, one who does not participate in the hunt but partakes of the spoils, and a dancing bee who finds nectar but does not alert other bees to its whereabouts, are defective. They are just as defective in these social aspects as are members of species who suffer from individual defects, for example, who lack sight, hearing, or mobility.

Foot (2002a, p. 200) claims: "I am therefore, quite seriously, likening the basis of moral evaluation to that of the evaluation of behaviour in animals". She stresses that human communication and reason complicate these evaluations, not least by creating a diversity of goods exceeding those available to plants and other animals in extent and complexity. That said, the passage just quoted expresses the central thesis of the Hursthouse/Foot/Thompson approach to ethical naturalism. Foot expands on this, as does Hursthouse. Both draw on an important paper by Thompson (1995), "The Representation of Life", which develops a conceptual structure for naturalistic evaluations of plants, animals, and humans.[7]

Thompson is interested in characterising representations and knowledge of things as alive, including the human life form. A starting point for understanding his view is his development of the abstract concept of a "natural kind" into the notion of a life-form. A life-form is the idea of a living kind or species. This abstract representation, which in the case of a particular species is built up partly from empirical observations of the lives of species members but go beyond them, is necessary if one is to understand both the individual members as well as the species itself. Knowledge of life forms enables us to make life-form attributions, the general form of which is: "*X is a bearer of life form S*, or *X is a member of species S*", and natural historical judgments (Thompson, 2004, p. 58, italics his). Natural historical judgments describe life-forms, taking the form: "The *S* is (or has, or does) *F*", as in " 'The domestic cat has four legs, two eyes, two ears, and guts in its belly' " (Thompson, 1995, p. 281). Thompson calls such sentences "Aristotelian categoricals", by which he means that they are statements of facts about species on which depends the good of individual species members. In more technical terms, natural-historical judgments provide the basis for "judgments of natural standard"; that is, species-relative judgments of natural goodness and natural defect. Natural-historical judgments are normative in allowing for inferences of the type: "The S is F", "this S is not F"; therefore, "this S is defective in not being an F". Through this type of inference, we arrive at judgments of the natural goodness or defectiveness of individual members of species.

This scheme applies to Foot's work insofar as operations of the will or practical reason in humans are species-dependent, and thus can be judged as good or bad according to a natural standard. Our species is the kind of life-form whose natural history shows us capable of acting well, for reasons that we see as good. Thus, when we fail to see reasons as good and do not act upon them, or see them as good yet fail to act, we are displaying defects, and if someone fails chronically in these ways, she can be judged a defective member of the human species. These judgments have the same conceptual structure and follow the same logic as attributions of goodness and defect to other species.

Ethical naturalism so conceived has come under fire from many scientifically minded critics.[8] The central issue, as Moosavi (2018, p. 296) puts it, is that ethical naturalism "relies on commonsense descriptions of living things rather than state-of-the-art science". This, she notes, is a problem of "naturalistic credentials" (Moosavi, 2018, p. 282). Thus, though ethical naturalism is "naturalistic" in the sense that it eschews Moorean or Platonic non-naturalism, as well as reliance on theistic sources of normativity (and non-cognitivist and Kantian approaches), it does not comport well with nature as scientists understand it.

Ethical naturalists seem unconcerned by this. Hursthouse (2004) argues that evolutionary biology does not undercut ethical naturalism, and that Foot was not trying to derive ethical judgments from a biological foundation – the project is coherentist, from within a certain viewpoint, not foundationalist, and does not try to derive ethics from biology. Instead, the project is about adjusting factual and ethical beliefs in a Neurathian manner from within a conceptual framework.[9]

According to Thompson (2008, p. 65), natural facts, which provide the basis for Aristotelian categoricals – that on which the good of a species and its members depends – can be "read off" such sources as nature documentaries and field guides. Millgram (2009, p. 559) calls this the "method of the nature documentary". It is the idea that identifying the natural facts about species that will count as Aristotelian categoricals can be considered an exercise in natural history. Clearly, defenders of ethical naturalism do not take what they are doing to be part of the scientific enterprise. Scientifically minded critics think this a mistake, and launch a barrage of criticisms against it. I will not pursue these, but simply register my belief that the critics are generally correct. Probing more deeply into the science of species and their histories – abandoning the method of the nature documentary for more sustained investigations into sciences and topics such as evolutionary biology and the mechanisms of heredity, need not take biological facts as foundational to ethics. Instead, doing so would enrich the conceptual framework of ethical naturalism with facts provided by contemporary science, and allow them to be put into sustained reflective equilibrium with our ethical beliefs.

This approach has "pros" and "cons". It could add much needed specificity to the vague naturalistic claims made by defenders of ethical naturalism, yet what bedevils the traditional picture is in the details. We can get a glimpse of this by noting that "nature" has long been altered in ways that support human interests. Long before the advent of germline genome editing, plant and animal species were altered to serve human interests through such techniques as selective breeding and, more recently, cloning. In the case of plants and animals, this could count against the naturalistic credentials of ethical naturalism, for the Aristotelian categoricals that we can identify for altered species are not entirely nature-based, but are based in part on human interests.

Selective breeding has also occurred in the human species in the form of eugenics. This has rightly been morally condemned, yet other, more benign examples can be given of the how the human choice of environmental factors influences heritable traits. Earlier, I mentioned that language and literacy produced populations that were more tolerant than their illiterate, more violent predecessors, and that heritable changes wrought through stable social structures, such as education and religion, produced changes in the kinds of goods and virtues constitutive of human flourishing. Pinker provides an example, writing:

> Throughout this book I have assumed that human nature, in the sense of the cognitive and emotional inventory of our species, has been constant over the ten-thousand-year window in which declines of violence are visible, and that all differences in behavior among societies have strictly environmental causes.
>
> (2011, p. 62)

That is, environmental interactions with basic human faculties such as language, reasoning, and emotion have not produced significant changes in natural capacities, but instead have altered expressions of traits and behaviours that have been inherited by successive generations and have tended toward the overall reduction of violence in the 10,000 years that he investigates. He gives examples of the Yanomamö and the Jivaro (Shuar) of Ecuador – men who killed other men had more wives and children than those who didn't kill. Had this trajectory continued over many generations, natural selection would have favoured a genetic tendency to kill. By contrast, societies that chose to be monogamous would have removed the reproductive advantage involved in killing. Natural selection in those societies would have favoured less violent tendencies.

Let us pause to make some observations. The first is that many scientists would not accede to Pinker's assumption that human faculties such as language, reasoning, and psychological and emotional capacities remained constant over the 10,000 years that he studied. As previously mentioned, manipulations of nature through technologies and cultural factors have produced changes in human capacities. Capacities might have changed slowly and gradually, but they did indeed change. We now face the possibility that human capacities such as language, reasoning, and psychological and emotional capacities, can be altered far more quickly – perhaps in a generation or two – through germline genome editing.

The second observation has to do with the role of human choice in producing heritable traits. This issue is not raised exclusively by germline genome editing, but is highlighted by it writ large. For ethical naturalists, evaluations of moral goodness are thought to be continuous with evaluations of natural goodness. In adopting this framework, Foot proposed to depart from the view that morality is based on, or intimately connected with, human interests. Yet the history of the human species seems to be pervaded by human choices, a large proportion of which, we can plausibly assume, were made to promote human interests, and did not proceed from or accord with any recognisably ethical outlook. We cannot assume that all or even most of the choices that shaped our evolutionary trajectory were made by possessors of practical wisdom who recognised the value of virtue and chose accordingly.[10]

My third point takes us back to the central question of this volume – whether virtues are local or universal. Even a brief glimpse into evolutionary theory suggests that traits and virtues develop evolutionarily in response to environmental contingencies. Human choice is always implicated in the development of virtues, and that choice depends on the circumstances in which people find themselves. There is a radical locality in the evolutionary development of virtues that leads to the suspicion that whatever universality we might see virtues possessing should be explained in historical and anthropological, rather than naturalistic, terms. For example, the choice of one society to invade, conquer, and impose its customs and mores on others is a way in which virtues are generalised across cultures; the choices of some societies to engage in practices that cause them to die out is a way in which some purely local virtues might have been eliminated. Defenders of neo-Aristotelianism might want to agree with Pinker's assumption and contend that what is stable and universal in this picture is human capacities, not virtues. Stability and similarity among human capacities would therefore account for the universality of virtues, whereas the environmental contingencies of localised development would explain variability. We should not and need not look to historical facts such as invasions and conquests to explain the universality of virtues, but instead to human capacities. But the assumption of the stability, if not the similarity, of human capacities is contested, not only in evolutionary theory, but also by the prospect of germline genome editing. Some background on germline genome editing will give a sense of the urgency of the questions it raises.

Genome editing in brief[11]

In 2017, the Committee on Human Gene Editing (CHGE) of the National Academies of Sciences, Engineering, and Medicine published a comprehensive report entitled *Human Genome Editing: Science, Ethics, and Governance*. The authors write: "Genome editing is a powerful new tool for making precise additions, deletions, and alterations to the genome – an organism's complete set of genetic material" (CHGE 2017, p. 15). They use the term "genome editing" instead of "gene editing" because "the editing could be targeted to sequences that are not parts of genes themselves, such as areas that regulate gene expression" (CHGE 2017, p. 15, n. 1). The committee writes that new approaches to genome editing, such as the use of meganucleases, zinc finger nucleases (ZFNs), transcription activator-like effector nucleases (TALENs), and most recently, the CRISPR/Cas9, make genome editing more precise, efficient, flexible, and less expensive to use than previous systems. Here I focus on CRISPR.[12]

Jennifer Doudna, of the University of California at Berkeley, is one of the scientists who pioneered the development of the CRISPR system. In an interesting and wide-ranging popular book chronicling its discovery, Doudna and her colleague Samuel H. Sternberg explain multiple ways in which CRISPR functions as a kind of "molecular scissors" to cut or edit gene sequences in various ways. They also explain how deactivating CRISPR provides a way for scientists not to permanently edit a genome's DNA, but to change "the way that its DNA gets interpreted, translated, and expressed" (Doudna & Sternberg, 2017, p. 108). Thus, gene expression control – turning genes on or off or dialing them up or down – is another function that can be performed by CRISPR, though in its deactivated form (Doudna & Sternberg, 2017, p. 109).

CRISPR has been used to edit the genes of plants, insects, and animals, but the most controversial of its uses are in humans. Two kinds of CRISPR interventions are possible: changes to somatic cells, such as muscle, heart, liver, etc., the DNA of which cannot be transmitted to offspring; and germ cell interventions. The DNA of germ cells is heritable. Germ cells "make up the germline of the organism – the stream of genetic material that is passed from one generation to the next" (Doudna & Sternberg, 2017, p. 158). Somatic and germ cell CRISPR interventions are of two kinds: those that seek to repair deficiencies, and those that seek to enhance human capabilities. To be sure, the repair of a deficiency can also be seen as an enhancement in a specific case, insofar as it brings an individual's capacity up to a baseline level. So, for example, if a Parkinson's sufferer has genome-editing therapy and because of that, the elimination of tremors, she could be said to be both repaired and enhanced.[13] However, in the literature on enhancement, "enhancement" is typically seen as raising individual capacities beyond normal human baselines.

Many scientists counsel caution.[14] As early as 2015, some scientists sought a ban on germline editing. In the same year, a group including Doudna and Sternberg published guidelines for moving forward. Harvard Medical School's Personal Genetics Education Project has launched a Genetics Consortium to educate people across the country, and from 27–29 November 2018, the Second International Summit on Genome Editing was held at the University of Hong Kong.

Yet on 26 November 2018, *The Wall Street Journal* reported that a Chinese scientist, He Jiankui, claimed to have produced the world's first genetically modified babies, twin girls (see Rana & Fan, 2018). The work, which was not vetted by other scientists and has not been published, apparently consisted of altering the genes of embryos to resist HIV infection. The article suggests that the embryos resulted from a healthy mother and an HIV-infected father, were genetically altered in a lab, then re-implanted in the mother's womb. Twin girls were born. More recently, the Chinese state news agency Xinhua reported that a third genetically modified baby was born and that He was convicted and jailed for three years. A Russian biologist has announced his intention to create more CRISPR-edited babies. These events have prompted the creation of a commission to propose criteria to guide clinical applications of human genome editing.[15]

Long before the furor over genome editing in embryos, Regalado (2016b, p. 82), raised a key ethical issue in the CRISPR debate: "The potential to precisely and easily 'edit' any genome using CRISPR is changing the way we think about nature". He also asked (Regalado 2016a, p. 57), "Does any country, agency, or individual have the right to change nature in ways that could affect everyone?" Changing nature includes changing human capacities, and also what could count as human goods and virtues.

Universality revisited and normative ethical issues

Might neo-Aristotelians have a way of meeting the heritability challenge to the traditional picture? The issue is not only that virtues are radically local, but also that human capacities have changed over evolutionary history. Nussbaum (1988) offers a framework that could address this. She observes that Aristotle thinks virtues should regulate human choices in various spheres of experience that can and do overlap and identifies eight such spheres: mortality, the body, pleasure and pain, cognitive capability, practical reason, early infant development, affiliation, and humor. How we interpret human experience within those spheres admits of cultural variation, and virtues and goods are also tempered and inflected with cultural meanings. The spheres themselves are sites for the exercise of human capacities. Even though human capacities have changed over evolutionary history, they have not changed in ways that are not recognisably human. In other words, even if we go back to our early evolutionary ancestors, we can still recognise their choices as falling within the ambit of the spheres, and we can identify them as human. Consider the sphere of bodily choices. Our ancestors did not have the choice of whether to exercise at the gym, but they had to care for their bodies – they had to seek food and drink in response to hunger and thirst, sleep and rest in response to fatigue, and, perhaps, take herbal remedies in response to illness. Like us, they were subject to the same kinds of physical injuries, and when those injuries were mortal, their lives ended. The point is that, even though human capacities have changed during the course of evolution, they have not changed in ways that we cannot acknowledge as species-dependent, and in some cases, such as the development of capacities for learning, literacy, art, and music, as species-unique.

Developing this line of thought, suggested here in barest outline, is a promising tack to take in light of the challenge that evolution poses to the stability of human capacities. The capacities, though not stable, are stable enough and similar enough over time to allow for goods and virtues to be grounded in recognisably human choices made in the context of the spheres of life that humans experience. Let us suppose for the sake of argument that this approach adequately addresses the evolutionary challenge to the stability of human capacities and allows neo-Aristotelians to ground the universality of virtues in facts about natural capacities. Germline genome editing threatens to upset this picture and raises numerous normative ethical questions.

Consider mortality. One good that germline genome editing could provide is increased longevity, especially for those whose genetic diseases, such as hypertrophic cardiomyopathy, can suddenly end life, or Duchenne's muscular dystrophy, whose sufferers typically do not live to reproduce. Are those who adopt the framework of ethical naturalism obligated to say that we must use germline genome editing to eradicate such diseases, which can be considered natural defects? Or would doing so disrupt our "characteristic way of going on", which ethical naturalists have arrived at through commonsense observation? Ethical naturalism gives no clear guidance, but we can say that among our natural endowments, one seems to stand out – the instinct for self-preservation. This is not exactly a capacity, but instead, a strong desire or drive to keep ourselves in existence. If we regard self-preservation as a rather strong and mostly constant drive or instinct in well-functioning members of the human species, it would seem that the choice to eliminate these diseases through germline genome editing has a firm natural basis, and is something that ethical naturalists should normatively endorse.

Yet there are more fine-grained questions. Recall the Chinese scientist, He Jiankui, who edited embryonic genes in order to make twins more resistant to HIV. As Greshko (2019) reports, "the edits may actually have hurt the babies' lifespans". Citing a study in the journal *Nature Medicine*, Greshko (2019) writes that, "researchers analysed a U.K. genetic database and found that when people naturally have a trait similar to the one that He engineered into the babies' DNA, they have about a 21% greater risk of dying before the age of 76 than people who don't have this trait". The upshot is that the lives of those who have no choice in the matter of whether they want their genomes edited could have their lives shortened. We should also note that genome editing, at this stage of its development, provides no guarantee that the quality of life of those individuals who choose to have their genes edited to eradicate disease, in themselves or their descendants, will be free of side effects. This is because the effects of the removal or alteration of human genes are not well known. We cannot yet say that germline genome editing promises to be a pure, unalloyed good in the sphere of human mortality. Can we say that those who choose to use this technology act virtuously, from the blameless desire to extend their own lives or those of loved ones? Perhaps, but the case could also be made that they are acting imprudently, before adequate knowledge of the effects of human germline genome editing is known.

Consider now the body. Genome editing has the capacity to directly affect the body, but also, through its effects on the body, it can affect the spheres of pleasure and pain, cognitive capability, practical reason, early infant development, and affiliation. As Nussbaum (1988) observes, bodies are much the same across cultures. Humans typically have the same general phenotype, the same dietary and other physical needs, the same approximate life spans, and the same range of physical capacities, no matter their race, ethnicity, or social and cultural situation. Germline genome editing has the capacity to change this in ways not evident in evolutionary history.

We raised questions about changes to the body in our discussion of mortality, but others present themselves. For example, if, through alteration of their genome, some people no longer have to worry about diet and exercise to stay healthy, their capacities for temperance, perseverance, and self-regulation could lapse, as these would no longer be necessary to maintain fitness. If some people find themselves immune to disease or illness, would their capacity for compassion for the suffering of others diminish? Would they lose the ability to empathise with those who have not received or do not wish to have genome alterations? If genome editing is costly and not widely available, would this foster attitudes of snobbery on the part of the recipients of health-enhancing or life-saving editing, and envy, resentment, and the feeling of having been unjustly denied a crucial good on the part of those unable to afford it?

The last question raises the issue of distributive justice, which directly affects the sphere of affiliation. Genome editing is still in early stages. Changes to goods at this point do not affect the species as a whole. Before they affect the species as a whole, they will affect the lives of individuals. It is worth asking how putative goods secured through genome alteration at the individual level will be distributed. Who will have access to genome editing, and on what basis? Will those who wish to secure the potential benefits of genome editing be selected on the basis of need? Presumably, sufferers of disease would be first in this queue. Will those in this group be selected on the basis of the severity of their disease, on their prospects for successful treatment, on the basis of their ability to pay, on the basis of some combination of these factors, or on some other basis entirely, such as their influence, power, affiliation with an "in group", and so on? If access to genome editing is not first made available to disease sufferers, on what other basis would it be made available? Wealth? Influence? Power? Connections? In such cases, there would be nothing to prevent the technology from being used first to create enhancements for those in privileged positions, and only later to be used to ameliorate the plight of disease sufferers. This scenario, in my opinion, would be an unjust distribution of a resource – one that would likely foment great resentment and social discord.

Let us pursue the question of enhancements. We know that germline genome editing offers the prospect of eliminating diseases, such as hypertrophic cardiomyopathy and hereditary blindness. Suppose that someone chooses to use this technology to alter herself and her descendants. This editing will have direct, and, let us suppose, largely benign, effects on her body, health, and capacities to experience pleasure and be free of pain, as well as those of her descendants. Why should she stop with simply eradicating a disease? Why not choose editing that will enhance her physical capacities? This might seem far-fetched, but consider the prospect of "designer babies" – whose genome is selected not only to eliminate natural defects, such as high risks of disease, but also to produce desirable traits, such as gender or eye color. Marcus (2018) discusses fertility clinics and services that use DNA testing to predict which embryos are at higher risks of certain diseases, such as diabetes or cardiovascular disease, to predict the gender of the embryo, and even to predict eye color.

The traits and capacities in question need not be only physical, but could eventually come to encompass morally relevant traits and capacities, such as those for reason, emotions, empathy, perseverance, and so on. Extensive debates about such changes and their implications for the human good can be found in the literature on the ethics of human enhancements. Buchanan (2011), for example, suggests that moral enhancements by biomedical means could improve our cognitive powers, our moral emotions, and our capacities for sympathy and moral imagination. The issue of enhancements bears on several of Nussbaum's spheres: pleasure and pain, cognitive capabilities, practical reason, and the nature of affiliation. A basic question is whether genetically engineered enhancements are truly goods that would foster virtues, or only apparent goods that would instead foment vice. Would the choice of enhancements be an exercise of practical reason? Would it truly increase pleasure, or instead bring about unforeseen pain or pleasure for the wrong reasons, such as vanity, expressed by the desire to have good looks, beautiful eye color, and so on? In the spheres of early infant development and affiliation, would parental love change if parents knew they could have a "designer" baby – one with perfect looks, health, a wonderful temperament, and desirable qualities – yet had to "settle" for an ordinary, natural baby without genetic enhancements?

How would the presence of genetically enhanced people affect the rest of the population? Would others view them as superior or as anomalies? Would the genetically enhanced view themselves as morally superior to the "unenhanced"? In other words, how would people interpret the experience of having enhanced, as opposed to unenhanced, capacities for sympathy, compassion, justice, and so on? Would enhanced capacities be considered "the new normal", and unenhanced capacities be considered subpar? Or would those with enhanced capacities be considered "supermoral" and subject to more stringent moral norms and expectations, as opposed to the "moral", who would have unenhanced capacities and be held to less stringent moral standards? How would the genetic modification of one morally relevant capacity or trait affect the possession of other, unmodified capacities and traits within a single individual's moral psychology? Could it have a detrimental effect on her usage of unenhanced capacities and traits, since, by comparison with her enhancement, the unenhanced might now seem defective?

Readers might be disappointed that I have raised more questions than I've answered. Surely, the metaethical questions raised by evolutionary theory, heritability theory, and germline genome editing for neo-Aristotelian ethical naturalism, their implications for the universality of virtues, and the possible reply based on Nussbaum's approach, cry out for further development. Normative ethical questions, too, should be put on the agenda for careful ethical consideration. Germline genome editing is at a crossroads. Many members of the scientific community now wish to impose a full moratorium on germline genome editing for clinical use. The time for deep ethical questioning is now.

Notes

1 See Robertson (2019) on *nunchi*.

2 I owe these insights to Patrick Frierson. I am grateful to him for helpful conversation and extensive comments on an earlier version of this essay. Thanks are also due to audiences at the Jubilee Centre Conference on "Virtues: Local or Universal", Oriel College, Oxford, UK, January 2019, and at Concordia University, Montreal, Canada, 8 November 2019. These interactions led to significant departures from my earlier approach. I also thank Catherine Darnell and Kristján Kristjánsson for their kind invitation to contribute to this volume, as well as for helpful comments.

3 I use the phrase "ethical naturalism" to refer specifically to neo-Aristotelian ethical naturalism as articulated by Hursthouse, Foot, and Thompson. In this section, I draw on but modify work initially presented in Snow (2018).

4 In addition to Geach (1977), Hursthouse, Foot, and Thompson, see McDowell (1998) and MacIntyre (1999).

5 See Foot (2002b, pp. 159–174, 2002a, pp. 189–208). References to Geach and Anscombe can be found in both papers; see Geach (1956, pp. 33–42), cited at Foot (2002b, p. 162), and Anscombe (1981) cited at Foot (2002b, p. 168), and Anscombe (1981, pp. 15, 18–19, 100–101, and 139), quoted and cited at Foot (2002a, pp. 198–199). See also Geach (1977, p. 17) quoted in Foot (2001, p. 35) and cited at p. 44.

6 For "Morality as a System of Hypothetical Imperatives", see Foot (1978); for an explanation of her departure from this view, see Hursthouse (2018).

7 Thompson has amplified his approach in later work (see Thompson, 2003, 2004, 2008, 2013).

8 See, for example, Millgram (2009), Moosavi (2018). For defenses, see Hursthouse (2012) and Lott (2012).

9 See also McDowell (1998).

10 Some critics of ethical naturalism press a variant of this point in a starker way by raising the so-called Pollyanna problem: what, in the history of human evolution, is continuous with human nature is not always morally good, and is sometimes morally evil. See Millgram (2009) and Moosavi (2018); for a reply, see Lott (2012).

11 This section draws on and updates unpublished material from Snow (2017), my inaugural lecture for the research series of the Margaret Beaufort Institute at Cambridge University, Cambridge, UK, "Proliferating Virtues: A Clear and Present Danger?" presented 27 October 2017.

12 A concise explanation of CRISPR is found on the website of the Broad Institute of MIT and Harvard (This Is Broad, n.d.), which researches genomics in order to improve human health (www.broadinstitute.org/what-broad/areas-focus/project-spotlight/questions-and-answers-about-crispr):

> "CRISPR" (pronounced "crisper") stands for Clustered Regularly Interspaced Short Palindromic Repeats, which are the hallmark of a bacterial defense system that forms the basis for CRISPR-Cas9 genome editing technology. In the field of genome engineering, the term "CRISPR" or "CRISPR-Cas9" is often used loosely to refer to the various CRISPR-Cas9 and -CPF1, (and other) systems that can be programmed to target specific stretches of genetic code and to edit DNA at precise locations, as well as for other purposes, such as for new diagnostic tools. With these systems, researchers can permanently modify genes in living cells and organisms and, in the future, may make it possible to correct mutations at precise locations in the human genome in order to treat genetic causes of disease.

13 This scenario is not as far-fetched as one might think. See Regalado (2016b).

14 See, e.g., Lanphier, Urnov, Haecker, Werner, and Smolenski (2015), and Baltimore et al. (2015).

15 For more on He, see Lo (2019); for the Russian scientist, see Cyranoski (2019); for more on the commission, see Lewis (2019).

References

Anscombe, G. E. M. (1981). *Collected philosophical papers* (Vol. 3). Minneapolis, MN: University of Minnesota Press.

Baltimore, D., et al. (2015). A prudent path forward for genomic engineering and germline gene modification. *Science*, *348*(6230), 36–38.

Buchanan, A. (2011). *Beyond humanity? The ethics of biomedical enhancement*. Oxford: Oxford University Press.

Committee on Human Genome Editing (2017). *Human genome editing: Science, ethics, and governance*. Washington, DC: The National Academies Press.

Cyranoski, D. (2019). Russian biologists plan more CRISPR-edited babies. *Nature*, *570*. Retrieved from www.nature.com/articles/d41586-019-01770-x?utm_campaign=Feed%3A+nature%2Frss%2Fcurrent+%28Nature+-+Issue%29&utm_medium=feed&utm_source=feedburner

Doudna, J. A., & Sternberg, S. H. (2017). *A crack in creation: Gene editing and the unthinkable power to control evolution*. New York: Houghton Mifflin Harcourt.

Foot, P. (1978). Morality as a system of hypothetical imperatives. In *Virtues and vices and other essays in moral philosophy* (pp. 157–173). Berkeley and Los Angeles, CA: University of California Press.

Foot, P. (2001). *Natural goodness*. Oxford: Clarendon Press.
Foot, P. (2002a). Does moral subjectivism rest on a mistake? In *Moral dilemmas and other topics in moral philosophy* (pp. 189–208). Oxford: Clarendon Press.
Foot, P. (2002b). Rationality and virtue. In *Moral dilemmas and other topics in moral philosophy* (pp. 159–174). Oxford: Clarendon Press.
Geach, P. (1956). Good and evil. *Analysis*, *17*(2), 33–42.
Geach, P. (1977). *The virtues*. Cambridge: Cambridge University Press.
Greshko, M. (2019, June 3). *First gene-edited babies may be at risk of early death*. National Geographic. Retrieved from www.nationalgeographic.com/science/2019/06/first-gene-edited-crispr-babies-may-be-risk-early-death/
Hursthouse, R. (2004). On the grounding of the virtues in human nature. In J. Szaif & M. Lutz Bachmann (Eds.), *Was ist das fur den Menschen Gute? What is good for a human being?* (pp. 263–275). Berlin, Germany: De Gruyter.
Hursthouse, R. (2012). Human nature and Aristotelian virtue ethics. *Royal Institute of Philosophy Supplement*, *70*, 169–188.
Hursthouse, R. (2018). The grammar of goodness in Foot's ethical naturalism. In J. Hacker-Wright (Ed.), *Philippa Foot on goodness and virtue* (pp. 25–46). Cham, Switzerland: Palgrave Macmillan.
Lanphier, E., Urnov, F., Haecker, S. E., Werner, M., & Smolenski, J. (2015). Don't edit the human germline. *Nature*, *519*(7544), 410–411.
Lewis, T. (2019, August 15). Scientists seeking better guidelines for editing genes in human embryos. *Scientific American*. Retrieved from www.scientificamerican.com/article/scientists-seek-better-guidelines-for-editing-genes-in-human-embryos/
Lo, K. (2019, December 30). China's gene-editing "Frankenstein" jailed for three years in modified baby case. *South China Morning Post*. Retrieved from www.scmp.com/news/china/science/article/3043894/chinas-gene-editing-frankenstein-jailed-3-years-modified-baby
Lott, M. (2012). Moral virtue as knowledge of human form. *Social Theory and Practice*, *38*(3), 407–431.
Macintyre, A. (1999). *Dependent rational animals*. Chicago and La Salle, IL: Open Court.
Marcus, A. D. (2018, October 2). Is it ethical to choose your baby's eye color? *The Wall Street Journal*. Retrieved from www.wsj.com/articles/is-it-ethical-to-choose-your-babys-eye-color-1538487936
McDowell, J. (1998). Two sorts of naturalism. In *Mind, value and reality* (pp. 167–197). Cambridge: Harvard University Press.
Millgram, E. (2009). Critical notice of *Life and Action*. *Analysis*, *69*(3), 557–564.
Moosavi, P. (2018). Neo-Aristotelian naturalism and the evolutionary objection: Rethinking the relevance of empirical science. In J. Hacker-Wright (Ed.), *Philippa foot on goodness and virtue* (pp. 277–308). Cham, Switzerland: Palgrave Macmillan.
Nussbaum, M. (1988). Non-relative virtues: An Aristotelian approach. *Midwest Studies in Philosophy*, *12*, 32–53.
Pinker, S. (2011). *The better angels of our nature*. New York: Penguin Books.
Questions and Answers about CRISPR (n.d.). Retrieved from www.broadinstitute.org/what-broad/areas-focus/project-spotlight/questions-and-answers-about-crispr
Rana, P., & Fan, W. (2018, November 16). Chinese scientist claims world's first genetically modified babies. *The Wall Street Journal*. Retrieved from https://

www.wsj.com/articles/chinese-scientist-claims-worlds-first-genetically-modified-babies-1543258316

Regalado, A. (2016a, April 13). The extinction invention. *MIT Technology Review*. Retrieved from www.technologyreview.com/s/601213/the-extinction-invention/

Regalado, A. (2016b, October 17). Can CRISPR save Ben Dupree? *MIT Technology Review*. Retrieved from www.technologyreview.com/s/602491/can-crispr-save-ben-dupree/

Robertson, S. (2019). *Nunchi*, ritual, and early Confucian ethics. *Dao*, *18*, 23–40.

Snow, N. E. (2017, October). *Proliferating virtues: A clear and present danger?* Paper presented at the Inaugural Margaret Beaufort Institute Research Series, Cambridge University.

Snow, N. E. (2018). Neo-Aristotelian virtue ethics. In N. Snow (Ed.), *The Oxford handbook of virtue* (pp. 321–342). New York: Oxford University Press.

This Is Broad (n.d.). Retrieved from www.broadinstitute.org/about-us

Thompson, M. (1995). The representation of life. In R. Hursthouse, G. Lawrence, & W. Quinn (Eds.), *Virtues and reasons: Philippa Foot and moral theory* (pp. 247–296). Oxford: Clarendon Press.

Thompson, M. (2003). *Three degrees of natural goodness*. Retrieved from www.pitt.edu/~mthompso/

Thompson, M. (2004). Apprehending human form. In A. O'Hear (Ed.), *Modern moral philosophy* (pp. 47–74). Cambridge: Cambridge University Press.

Thompson, M. (2008). *Life and action: Elementary structures of practice and practical thought*. Cambridge: Harvard University Press.

Thompson, M. (2013). Forms of nature. In G. Hindrichs & H. Axel (Eds.), *Freiheit* (pp. 701–735). Stuttgarter Hegel-Kongres, Frankfurt and Main: Vittorio Klostermann.

Section 2
Psychological issues

Nestled at the centre of our volume, Section 2 reflects contributions that approached the question of universality or locality of virtues by drawing on psychology theory and empirical research. Whilst these chapters offer a psychological perspective, none of these contributions can be accused of understanding virtues in a psychological vacuum, with each chapter demonstrating and exploring the important interplay between psychology, philosophy and education. All three chapters make some reference to virtue ethics, but each offers a slightly different conception of how this can be understood and linked to the behavioural demonstration of virtue. These differences are not surprising, given the authors explain virtues and moral behaviour using alternative psychological theories drawn from the behavioural and social sciences, ranging from motivational and trait theories in Chapter 7, to developmental systems in Chapter 8 and social domain theory in Chapter 9. Yet, despite these differences in theoretical approaches and the empirical evidence used to define and understand virtue, there is a consensus (as in Section 1 of this volume), that viewing universal and local virtues as a dichotomy is at the very least oversimplistic. Albeit to varying degrees, the chapters appear to emphasise a universality lens in relation to understanding and defining virtue, but when it comes to the demonstration of virtue, a locality lens which emphasises context, is seen as more appropriate. As such, these chapters provide answers to a more nuanced version of the "or" question, along the lines of "*When* are they local or universal?" and perhaps more importantly (as this is the psychology section), "*How* and *why* does this occur?"

In Chapter 7, Jayawickreme, Prentice, and Fleeson make a compelling case for morality as a basic psychological need. They argue that morality and virtues have universal importance, as the experience of "being moral" is not just something which is desirable but is in fact "needed". They discuss recent empirical research, which shows that the need to "feel moral" functions on a similar level to the basic psychological needs, identified by self-determination theory (Deci & Ryan, 2000), of autonomy, competence, and relatedness (ACR) and suggest that the "moral need" may act as a motivational force for moral character and the demonstration of virtue.

Jayawickreme, Prentice, and Fleeson suggest whole trait theory (WHT e.g., Fleeson & Jayawickreme, 2015) provides a useful framework for exploring this relationship, though they acknowledge that such a perspective aligns more happily with a virtue ethics conceptualisation of morality, as on a WHT account moral traits can be attributed to an individual without "absolute consistency" if occasional deviations from the trait are relevant to the context. Whilst it is not yet clear how the "moral need" may act as a mechanism for the manifestation of virtues, the chapter concludes by offering an array of fruitful directions for interdisciplinary research, setting out a tangible pathway to integrate motivation theory, personality and virtue ethics in the quest for virtue understanding.

In Chapter 8, Murray and Tirrell frame the understanding of virtue within a relational developmental systems (RDS) metatheory, which emphasises the reciprocal relationships between persons and contexts. Whilst they suggest this theory allows for a universal frame to define virtue, they also call for the integration of RDS with the specificity principle, by using developmental components of *phronesis* (Darnell, Gulliford, Kristjánsson, & Paris, 2019), to account for the "context-specific applications of virtue". In this sense, Murray and Tirrell argue for an understanding of virtue that both includes universality and locality, and they offer a theoretical account which would allow for the demonstration of this in an individual. Crucial to this integration is their definition of "coherence", which the authors distinguish from a coherence of character (on a trait-based account), to a coherence of action, which accounts for situational variation in virtuous behaviours. Using the virtues of humility and forgiveness as examples, the authors provide two helpful and convincing demonstrations of how the integration of RDS and phronesis could incorporate the universal and local aspects of virtue. The chapter concludes by calling for an approach to understanding virtue which emphasises the dynamic relationship between an individual and their context, making the case that such an approach could be used for the development of knowledge not only in the developmental sciences, but character education and philosophy, too.

In Chapter 9, Cole-Wright explores how we develop shared norms in socio-cultural groups and when change in relation to these norms might be considered "diversity" or "defiance". Using social domain theory (e.g., Nucci, 1981), Cole-Wright describes empirical evidence which shows that violations of norms, considered "moral" are more likely to be considered deviant than violations categorised as "personal" or "social/conventional". However, research also shows that this labelling can be changed based on whether individuals grounded these moral beliefs "objectively" or "non-objectively". Cole-Wright suggests these findings show an important role for meta-ethical pluralism in psychosocial functioning, as grounding the moral belief non-objectively (albeit temporarily) allows for further discussion about the belief, allowing for the possibility of diversity. Cole-Wright argues that virtues play a central role in these psychosocial processes, as virtues are what it "looks like to be moral". Using the virtue of respect as an example, she argues that the observation of what is it to be respectful and the experience of being respectful offers a feedback loop in the development of virtue (which also serves to safeguard socio-cultural space). The chapter ends by neatly drawing the reader back to the local or universal virtues debate, suggesting that shared norms will have an impact on the types of virtues prioritised within our social groups, giving space for variation both within and across cultures in relation to the demonstration of virtues.

References

Darnell, C., Gulliford, L., Kristjánsson, K., & Paris, P. (2019). Phronesis and the knowledge- action gap in moral psychology and moral education: A new synthesis? *Human Development*, *62*, 1–29.

Deci, E. L., & Ryan, R. M. (2000). The "what" and "why" of goal pursuits: Human needs and the self-determination of behaviour. *Psychological Inquiry*, *11*(4), 227–268.

Fleeson, W., & Jayawickreme, E. (2015). Whole trait theory. *Journal of Research in Personality*, *56*, 82–92.

Nucci, L. (1981). Conceptions of personal issues: A domain distinct from moral or societal concepts. *Child Development*, *52*, 114–121.

7 Morality as a basic psychological need

Preliminary evidence

Eranda Jayawickreme, Mike Prentice, and William Fleeson

Introduction

The purpose of this chapter is to review the evidence for morality functioning as a basic psychological need. We define the moral need as the experience that one is moral (Prentice et al., 2019b). Experiencing oneself as moral is a psychological nutrient necessary for an individual's integrity, growth, and adjustment (Vansteenkiste, Ryan, & Soenens, 2020). Thus, on our view, people require or need that experience. But does such a basic psychological need exist? Is the experience of greater moral need satisfaction tied to meaningful and impactful life events? How does moral need satisfaction predict moral behaviour in daily life? Does the relationship between moral need satisfaction and moral behaviour differ in people of high vs. low moral character? We believe that the question of whether morality functions as a basic psychological need (and by extension promotes flourishing) is central to the theme of this volume. This is because this research programme highlights on the universal importance of morality and virtue as something that is not just desirable but is *needed* (Ryan, Curren, & Deci, 2013).

This chapter summarises the conceptual rationale for the moral need against current criteria (e.g., Baumeister & Leary, 1995; Ryan & Deci, 2017). It reviews a set of studies (Prentice et al., 2019b; Prentice, Jayawickreme, & Fleeson, 2020) providing initial empirical support for morality as a fundamental psychological need. It considers how these findings can help integrate prominent theories of motivation (self-determination theory, SDT) and personality (whole trait theory, WTT) in the common pursuit of studying moral character traits (Jayawickreme, Zachry, & Fleeson, 2019; Prentice, Jayawickreme, & Fleeson, 2019a; Snow, Wright, & Warren, 2019). The chapter concludes with future directions for examining how the moral need may function as a key mechanism to promote moral character.

Is the need to feel moral a basic psychological need?

While our empirical work is the first to examine the possibility of morality functioning as a basic need (although see Martela & Ryan, 2016, 2020, for a related line of research), the idea that morality is vital for human flourishing has been argued throughout the history of psychology. For example, James (1878, p. 7) noted that "the joy of moral self-approbation . . . [may be] required to make the notion of mere existence tolerable". Building on this insight, Steele's (1988) self-affirmation theory claimed that people are motivated to believe that they are good (Prentice et al., 2019b). However, with the recent increase of interest in the empirical study of moral character (Fleeson, Furr, Jayawickreme, Meindl, & Helzer, 2014), there has been increased focus in psychology on the study of moral traits (e.g., Talevich, Read, Walsh, Iyer, & Chopra, 2017; Zeinoun, Daouk-Oyry, Choueiri, & van de Vijver, 2017), as well as on the role of character and morality for flourishing (Kristjánsson, 2020).

A first step in investigating whether morality is a basic psychological need is to discover whether the experience that one is moral functions in a manner similar to that found in empirical research for other basic needs (Vansteenkiste et al., 2020). To provide one instantiation of this question, does the need to feel moral function like the psychological needs of autonomy, competence, and relatedness (ACR) identified by self-determination theory (SDT; Deci & Ryan, 2000)? SDT defines a basic psychological need as "a psychological nutrient that is essential for individuals' adjustment, integrity, and growth" (Vansteenkiste et al., 2020, p. 1; see also Ryan, 1995). In line with basic need satisfaction theory (BNST), a subtheory of SDT, the satisfaction of these basic needs is essential to the achievement of well-being, while thwarting of them is associated with sub-optimal functioning (Deci & Ryan, 2000). A voluminous empirical literature has provided support for the key roles that autonomy, competence, and relatedness play in promoting human flourishing; yet the question of whether morality functions in a similar manner has remained unanswered up to now.

This is surprising, given that past research appears to have taken a moral need for granted. Past research appears to have often assumed the moral need motivates morally relevant behaviours (Prentice et al., 2019b). To provide an example, one line of research has found evidence for people being *less* likely to engage in moral behaviour when their own morality is highlighted, implying that their moral need has already been satisfied (Blanken, van de Ven, & Zeelenberg, 2015). Furthermore, recent work has highlighted the importance of moral character as a distinct dimension in the perception of others (Goodwin, 2015). Overall, such findings highlight the likelihood that morality may function as a basic psychological need.

We additionally believe that at least four reasons can be posited as justifications for the existence of such a need (Prentice et al., 2020). These justifications draw on work on the evolution of human collaboration, current research person perception, a logical analysis of human cognition and development, and our own preliminary empirical investigations.

Evolutional justification

First, work focused on evolutionary approaches to understanding human and primate morality suggests that moral behaviour may have a long phylogenetic history and that it may have facilitated the advanced social coordination that characterises human behaviour (e.g., de Waal, 2009; Tomasello, 2009). Given that the "inherent" nature of the need is a key criterion for BNST (Vansteenkiste et al., 2020), we believe that the evidence supports the view that morality meets this criterion and highlights the universality of questions of morality and character for flourishing.

The importance of morality for person perception

Recent work by Goodwin and colleagues suggests that moral character is likely the predominant dimension that is evaluated in person perception (Goodwin, 2015; Goodwin, Piazza, & Rozin, 2014). Furthermore, people can agree on moral character judgments, and such judgments are key to being liked, respected, and understood by others (Hartley et al., 2016; Helzer et al., 2014), as well as identifying "do-gooders" (Pleasant & Barclay, 2018). The key dimensions of person perception – warmth and competence (Fiske, Cuddy, & Glick, 2007) have analogical basic needs (relatedness and competence, respectively). It therefore follows that if morality is a key dimension in person perception, a psychological need should correspond to it (Prentice et al., 2019b).

Evaluative function of the moral need

We believe that there is a logical argument for morality functioning as a basic need. This argument is based on developmental or ontogenetic processes by which people adopt standards (Prentice et al., 2019b). Whatever the highest standard used to evaluate people's actions, most people are likely to have a need to be judged favourably on that standard. If this is the case, then if claims regarding morality represent the highest standard, it follows that there would be a need to be judged favourably on those moral standards. This argument may explain why people indicate that morality is the most essential part of their identities, selves, and souls, more so than other personality traits or their autobiographical memories (Strohminger & Nichols, 2014).

Summary of research on morality as a basic need

Baumeister and Leary (1995) provided a set of criteria for evaluating candidate psychological needs (see Table 7.1, reproduced from Prentice et al., 2019b). In light of existing research, we posited that morality already meets

Table 7.1 Criteria for establishing psychological needs from Baumeister and Leary (1995), the criteria that Prentice et al. (2019b) tested, and the cumulative passing of criteria for morality

Criteria for need	*Previous evidence*	*Tested*	*Passed*	*Cumulative*
1. produce effects readily under all but adverse conditions (prolific)	Goodwin (2015); Helzer et al. (2014)	X	X	X
2. have affective consequences (affective)	Tetlock et al. (2000)	X	X	X
3. direct cognitive processing (cognitive)	ambiguous Epley and Dunning (2000)	X	X	X
4. lead to ill effects (such as on health or adjustment) when thwarted (degenerative)	Guillen-Royo and Kasser (2015)			X
5. elicit goal-oriented behaviour designed to satisfy it (motivational)	Tetlock et al. (2000)			X
6. be universal in the sense of applying to all people (universal)		X	X	X
7. not be derivative of other motives (non-derivative)		X	X	X
8. affect a broad variety of behaviours (expansive)	Tetlock et al. (2000)			X
9. have implications that go beyond immediate psychological functioning (enduring)	ambiguous (cf., e.g., Helzer et al., 2014)	X	X	X

Source: Reproduced from Prentice et al. (2019b)

some of these criteria. Moral motivation has affective consequences (Tetlock, Kristel, Elson, Green, & Lerner, 2000; Rothschild & Keefer, 2017) and elicits morally relevant behaviour, thus satisfying the affective and motivational criteria (Prentice et al., 2019b).

However, other criteria lack direct empirical support. For example, despite Steele's (1988) proposal that people's desire to see themselves as morally adequate drives self-affirmation processes, the moral need has not been tested as a specific motivator of self-affirmation. Additionally, while morality is a key evaluative character trait (Goodwin, 2015; Helzer et al., 2014), such work examines perceptions of moral traits in others, rather than self-perceptions of moral adequacy. This raises the question of whether it is in fact reasonable to suspect that morality functions as a basic psychological need.

These considerations led to an initial set of empirical studies to examine the nature of morality as a basic need. In the following section, we present an overview of three studies – utilising five samples – that provide initial evidence that the need to feel moral does function in a manner consistent with basic psychological needs.

"Entrance exam"

In our first set of studies (Prentice et al., 2019b), we conducted an empirical "entrance exam" on morality using a method utilised by Sheldon, Elliot, Kim, and Kasser (2001) to provide evidence that ACR function as basic psychological needs. In two studies and four samples, participants recalled events in which they felt: un/satisfied, meaningful, pleasurable, at their best, and at their worst. They then rated the degree to which candidate psychological needs were satisfied during those events. Morality was frequently as much or more satisfied than ACR during events related to satisfied, meaningful, pleasurable, and "at their best" events.

Further, it was positively related to indices of positive functioning. Across both sets of studies, participants provided information on both their current satisfaction with life and psychological flourishing (Diener, Emmons, Larsen, & Griffin, 1985; Diener et al., 2010), as well as retrospective reports of their positive and negative affect and quality of life during important life events. Moral need satisfaction aggregated across the life events was positively correlated with flourishing (*rMTurk* = .67, *rstudents* = .40) and satisfaction with life (SWL; *rMTurk* = .50, *rstudents* = .24), as were the ACR needs and with comparable strengths. These findings for SWL and flourishing were replicated in another study with two samples in Study 2.

A series of regressions predicting SWL and flourishing from the needs revealed that morality and the ACR needs were all *unique*, positive predictors of both SWL and flourishing. Reduced positive affect and quality of life was observed for events in which the satisfaction of the moral need was low. The moral need was also shown to be one of the most responsive needs to the satisfying/thwarting dynamic across positive/negative life events.

These initial results confirmed that morality meets many of the criteria posited by Baumeister and Leary (1995). The need to feel moral was found to direct cognitive processing (in that participants utilised the need to identify peak experiences), have affective consequences, and to be distinctive from ACR. In other words, there was evidence in favour of the positive connection between moral need satisfaction and psychological well-being, and thus for a moral need to meet the primary criterion.

Examining the moral need in daily life

We next conducted an experience sampling study of momentary moral need satisfaction (Prentice et al., 2020). The goal of this study was to examine satisfying the moral need associated with the manifestation of moral behaviours and increased well-being. We also examined whether the moral need functions differently in people who rated themselves as being either high or low in moral character. The initial particular pool consisted of 295 Amazon Mechanical Turk workers drawn from a larger sample of 10,000 workers who had previously completed a set of measures assessing moral character (see Prentice et al., 2020 for more details). The participants were made up of three groups: the morally exceptional (top 5% of moral character scorers in the sample of 10,000 workers), the morally average (middle 5%), and the morally indifferent (bottom 5%). We refer the reader to Prentice et al. (2020) for a detailed justification for this extreme groups approach. These participants provided a total of 3,276 reports over the course of seven days, with six assessments per day (and an average of 32 reports per participant).

Satisfying the moral need was correlated with psychological thriving (assessed using items from the Brief Inventory of Thriving; Su, Tay, & Diener, 2014) both cross-sectionally and concurrently in everyday life, $r = .60, p < .001$. Multilevel analyses were carried out to investigate whether moral behaviour predicted moral need satisfaction and found a significant effect allowing for random intercepts, b = .43, SE = .02, t(3151) = 2.86, p < .001. Additionally, moral need satisfaction was the most tightly linked need to moral behaviour, demonstrating the specificity of content for the need. Specifically, moral need satisfaction remained a significant predictor of moral behaviour even after controlling for the effects of ACR; b = .21, SE = .03, t(3135) = 6.58, p < .001.

Additionally, participants who self-reported high levels of moral character reported more moral behaviour ($M_{BxHigh} = 5.02$, SD = 1.22) and moral need satisfaction ($M_{NeedHigh} = 5.95$, SD = 1.00; Fs ≥ 15.00, ps < .001) than those who reported average and low levels of self-reported character. Intriguingly, people who self-reported the lowest levels of moral character benefited the most from enacting moral behaviour and fulfilling their moral need, perhaps reflecting a desire for such experiences. Conversely, it is possible that people who report high levels of moral character may be less affected by the enactment of moral behaviour in their daily life due to either chronic satiation of the need or complementary satiation of other basic needs.

This study provided evidence for satisfaction of the moral need being related to psychological thriving in daily life over and above the psychological needs for autonomy, competence, and relatedness. Moral satisfaction is also found to be linked to moral behaviour, satisfying a key criterion for a basic psychological need (Ryan & Deci, 2017). However, we are limited in the causal claims we can make for this study, as the significant results found assessed moral need satisfaction and moral behaviour simultaneously. Future research needs to test the causal dynamics of moral need satisfaction, moral behaviour and flourishing, ideally in an experimental design (Prentice et al., 2020).

Integrating self-determination theory and research on the moral need

While we believe that morality likely constitutes a basic need based on the criteria provided by Baumeister and Leary (1995), it is less clear whether it meets the criteria to join ACR as a basic need as posited by SDT. Ryan and Deci (2017) have provided six criteria required of a candidate need to join ACR as a basic need within SDT. How does the moral need fare against these criteria? In Prentice et al. (2020), we provided a detailed evaluation of the extent to which morality is likely to meet these six criteria. We summarise that discussion following.

Psychological thriving

A key criterion is that a candidate need promotes psychological thriving. As Ryan and Deci (2017, p. 250) have noted, "a new candidate (need to) be strongly positively associated with psychological integrity, health, and well-being, *and* that its frustration be negatively associated with these outcomes, over and above the variance accounted for by the existing needs" (emphasis by the author). In our view, the evidence from our first two studies provide compelling initial evidence that the moral need does in fact promote thriving.

Specificity of content

This second criterion is that a need:

> must specify content – that is the specific experiences and behaviours that will lead to well-being. [ACR] make clear what people need to do in order to be healthy – for example, do important activities well, endorse their actions, and connect with others.
>
> (Ryan & Deci, 2017, p. 251)

In line with our definition of the moral need, the "content" of the moral need are the specific instances of meeting personal moral standards, usually by performing actions seen as moral that lead to experiences that one is morally good. We believe that examining whether moral need satisfaction is most frequently associated with specific moral behaviours (e.g., behaviours associated with fairness or beneficence; Weinstein & Ryan, 2010) or specific moral foundations (Graham, Haidt, & Nosek, 2009) represents important avenues for research on understanding the link between manifestations of specific moral character traits and satisfaction of the moral need.

Explanatory utility

The third criterion is that a candidate need be *essential* to explain or interpret psychological phenomena (Ryan & Deci, 2017, p. 251). In our view, the moral need may play an essential role in explaining moral motivation and behaviour. None of ACR focus specifically about meeting moral standards (Prentice et al., 2020). While we acknowledge that combinations of SDT components have been posited to understand instances of moral and immoral behaviour (e.g., Ntoumanis & Standage, 2009; Sheldon, Sommet, Corcoran, & Elliot, 2018), the moral need will likely provide a more parsimonious explanation of these effects.

Growth vs. deficit needs

Any new need should:

> be consistent with the idea of a growth need rather than a deficit need . . . there are two types of psychological needs that could be basic: growth needs that facilitate healthy development and are active on an ongoing bases and deficit needs that operate only when the organism has been threatened or thwarted.
>
> (Ryan & Deci, 2017, p. 251)

In our view, the preliminary studies presented in Prentice et al. (2019b) provide evidence for morality as a growth need in a number of ways. First, when people remembered peak experiences, morality was among the most satisfied needs during those episodes, indicating that moral satisfaction acted, in part, as a signal that life was going well during those times. Second, morality was linked to indices of flourishing. We have also noted that when people are prevented from making further progress on life aspirations that include other-regarding moral content (such as caring for one's community), their psychological development is impeded (Guillen-Royo & Kasser, 2015). This suggests a moral need that is constantly operative and that when people are prevented from hitting their targets, it leads to suffering.

Categorical appropriateness

Ryan and Deci (2017, p. 252) caution that candidate needs should not be conflated with the outcomes to be predicted from the satisfaction of basic psychological needs. To reference one example (Prentice et al., 2020), psychological health constitutes the outcome variables that are increased or decreased by need satisfaction. Thus, indices of psychological health, such as self-esteem or meaning, are outcomes, not needs. We think that moral need satisfaction should be examined as a predictor of flourishing, rather than as an indicator of well-being. That said, it could be argued that the moral need as measured is in fact confounded with self-esteem. This is because self-esteem in part constitutes meeting standards of social value. We do not think this is the case, as we believe that people possess the need to be moral by their standards, and furthermore that moral need satisfaction likely requires the manifestation of specific actions with specific motives and beliefs. However, the challenge in obtaining an objective judgment of actions associated with moral need satisfaction means that a subjective evaluation is the most optimal way to assess the need, as we have done in our empirical research.

Universality

Finally, Ryan and Deci (p. 252) argue that a basic need must apply universally – for all people in all ages in all cultures. Further, the need must be a reasonable fit to evolutionary considerations and convey advantages that would have driven its universality. We were motivated by similar concerns when approaching this question, as noted previously in our rationale for examining the moral need. In our own work (Prentice et al., 2020), we found that the satisfaction of the moral need does not appear to have differential impact on markers of well-being, even when trait-level preferences for moral experiences are taken into account, which underlines the notion that everybody benefits from moral need satisfaction, even if they might differ in the extent to which they seek out such experiences. Other evidence is found in the developmental psychology literature, which demonstrates that capacities for collaboration and helping appear to come online at a very early age (Tomasello & Vaish, 2013). Children have been found to keep promises and expect them to be kept by others, punish others for moral transgressions, and engage in restorative behaviour for their own transgressions (Hepach, Vaish, & Tomasello, 2017a; Kanngiesser, Köymen, & Tomasello, 2017; Mammen, Köymen, & Tomasello, 2018). Moral behaviour additionally appears to be intrinsically motivated; child transgressors exhibited distressed psychophysiology before engaging in reparative actions, and also exhibited positive mood when successfully helping others achieve their goals to a similar level as theirs (Hepach et al., 2017a; Hepach, Vaish, & Tomasello, 2017b).

In summary, the moral need appears to meet each of the six criteria for a basic psychological need based on initial evidence. While there are many important questions to examine, the available evidence suggests that the moral need may in fact meet the criteria posited by SDT. That said, even if subsequent research reveals that it does not meet criteria for a basic need *as defined by SDT*, it may still meet the criteria of other approaches to basic needs (e.g., Dweck, 2017).

The moral need as a mechanism for moral behaviour: extending whole-trait theory

Regardless, further research is warranted from researchers from motivational and personality traditions on the nature of the moral need due to the important intrapersonal/interpersonal implications of moral motivation so far uncovered (Fleeson et al., 2014; Meindl, Jayawickreme, Furr, & Fleeson, 2015; Jayawickreme, Meindl, Helzer, Furr, & Fleeson, 2014). Regardless of whether the moral need is incorporated into SDT, it is possible that this need functions as a key motivational mechanism for the manifestation of moral character, as posited by whole trait theory (Fleeson & Jayawickreme, 2015; Jayawickreme & Fleeson, 2017; Snow et al., 2019).

We have previously discussed the applicability of WTT to moral traits (see Jayawickreme et al., 2019; Jayawickreme et al., 2014; Jayawickreme & Fleeson, 2017). WTT is founded on Fleeson's (2001) seminal insight that personality traits are best understood as density distributions (patterns of behaviour that that vary widely, but around a consistent mean). This view has many advantages (Fleeson, 2004), among them the fact that this view of meaningful variability is consistent with widely held intuitions about most moral traits. WTT builds on this view of traits by giving separate standing to the descriptive aspect of traits (understood as density distributions of momentary state enactments) and explanatory aspects of traits, which constitute the motivational and social-cognitive mechanisms that give rise to the distributions. Together, these two sides have a distinctive standing as components of a *whole trait*.

Perhaps the key point of complementarity between WTT and SDT (Prentice et al., 2020) is that WTT suggests that traits vary in their enactments in the service of goals, and goals will often be ultimately aimed at the satisfaction of basic psychological needs (see also Sheldon, 2011). On this account, it is possible that goals that have the ultimate aim of promoting moral need satisfaction would lead to the enactment of morally relevant states.

For example, consider the virtue of honesty. Most people likely manifest different levels of honesty at different points in their daily life. However, it is likely that more honest people tend to be repeatedly more honest than less honest people are. Our view (Jayawickreme & Fleeson, 2017) is that a person with a higher mean level of honesty is more honest in an important way than is a person with a lower mean level of honesty. We also acknowledge that specific and heinous dishonest actions may affect judgments about a person (Jayawickreme & Fleeson, 2017). We would nevertheless be willing to tolerate occasional deviations from habitual honest behaviours in relevant contexts and, continue to see someone as honest. In other words, we believe that it is possible to attribute moral traits and vices to individuals without the need for absolute consistency. This is in contrast to a view that would simply view both individuals simply lacking virtue on a threshold perspective (Fleeson & Jayawickreme, 2017).

Prentice et al. (2020) provided suggestive evidence for thinking about the moral need from a WTT perspective by focusing on the manifestation of moral behaviour in daily life (see previously in this chapter). We believe that research designs such as these provide a valuable template for future research. Despite this initial evidence, however, we acknowledge that WTT provides a framework for thinking about the relationship between moral need satisfaction and the manifestation of specific moral virtues that is most amenable to a virtue ethics conception of morality (Snow et al., 2019). While we believe that a virtue ethics conceptualisation of morality is consistent with our conception of traits, we of course acknowledge that this is not the sole way in which the moral need may affect moral behaviour.

Nevertheless, thinking about moral traits from a WTT perspective affords a framework for examining the various mechanisms undergirding moral thoughts, feelings, and behaviours, and affords the opportunity to provide an explanatory account of many moral traits. One question that merits further work from both psychologists and philosophers is whether the need to feel moral is a necessary precursor for living well. Curren (2013) has posited that ACR may be *considered felt Aristotelian necessities*, or needs that must be fulfilled for individuals to reach their full potential. Assuming that future research confirms that the moral need does in fact meet SDT's criteria for a basic need, it may be that people do need to live their lives in accordance with moral standards in order to live well. The status of basic needs as a mechanism remains unclear (e.g., Snow et al., 2019), but future interdisciplinary work can clarify the role that the moral need plays in the manifestation of specific virtues.

Future directions and conclusion

Our initial studies have found that the extent to which people experience moral need satisfaction predicts both their flourishing and moral behaviour. From the perspective of SDT, future research should examine whether specific moral demands may negatively impact the satisfaction of ACR. One possibility in these contexts is that fulfilling one's moral need may sometimes be linked the thwarting of other basic psychological needs. This may be especially the case in instances when external or introjected moral demands impact people's behaviour. Future research should therefore examine the dynamic interplay between morality and ACR.

We note with interest that benevolence (the sense of having a positive impact in the lives of other people) has been recently tested as a candidate basic need (Martela & Ryan, 2016; 2020). Benevolence is clearly relevant to morality, and future research should examine the extent to which these two candidate needs are in fact distinct from each other. It should be noted, however, that Martela and Ryan (2020) acknowledge the possibility that not all benevolent actions may count as moral (Batson, Klein, Highberger, & Shaw, 1995).

It is also important to examine the distinction between the subjective experience of being moral (which is how the moral need was assessed, similar to assessments of ACR) and the act of being moral (i.e., doing the right action as a result of morally relevant motivations). We predict that the moral need will track real morally relevant behaviours in most contexts, but we acknowledge that it is possible that people may not maximise every opportunity to be moral (e.g., use moral "wriggle-room"; Mazar, Amir, & Ariely, 2008). Examining whether moral need fulfilment motivates moral behaviour may be a fruitful topic for researchers working in character education (Prentice et al., 2019b). Finally, research on which specific moral virtues best predict moral need satisfaction, and whether these virtues may vary across cultures, represents an important future direction.

In conclusion, we note that similar to WTT's recent interdisciplinary engagement with ethics (Jayawickreme & Fleeson, 2017; Fleeson & Jayawickreme, 2017), SDT researchers have begun exploring the interplay between psychological theories of flourishing and Aristotelian ethics (Curren, 2013; Kristjánsson, 2020; Ryan et al., 2013). We think that research on the moral need can provide philosophers, psychologists, and educators with a universal language to think about the vital role morality plays in promoting character and flourishing in different contexts (e.g., Pierrakos et al., 2019). Showing that morality functions as a basic need also provides support for the humanistic and Aristotelian view that people are amenable to being motivated by self-transcendent and moral motivations, and that satisfying such motivations is a key pathway to flourishing.

Acknowledgements

This chapter reproduces and summarises ideas and data first presented in Prentice et al. (2019b), Prentice et al. (2019a), and Prentice et al. (2020). Special thanks to Anna Hartley, Ashley Hawkins, Mike Furr, and Christian Miller for their work and insights that contributed to this research programme.

Correspondence concerning this chapter can be addressed to Eranda Jayawickreme, Department of Psychology, Wake Forest University, PO Box 7778, Winston-Salem, NC 27104. Email: jayawide@wfu.edu.

This publication was supported by grants from the Templeton Religion Trust and the John Templeton Foundation. The opinions expressed in this publication are those of the authors and do not necessarily reflect the views of either the Templeton Religion Trust or the John Templeton Foundation.

References

Batson, C. D., Klein, T. R., Highberger, L., & Shaw, L. L. (1995). Immorality from empathy-induced altruism: When compassion and justice conflict. *Journal of Personality and Social Psychology*, *68*, 1042–1054.

Baumeister, R. F., & Leary, M. R. (1995). The need to belong: Desire for interpersonal attachments as a fundamental human motivation. *Psychological Bulletin*, *117*, 497–529.

Blanken, I., van de Ven, N., & Zeelenberg, M. (2015). A meta-analytic review of moral licensing. *Personality and Social Psychology Bulletin*, *41*(4), 540–558. https://doi.org/10.1177/0146167215572134

Curren, R. (2013). Aristotelian necessities. *The Good Society*, *22*(2), 247–263.

Deci, E. L., & Ryan, R. M. (2000). The "what" and "why" of goal pursuits: Human needs and the self-determination of behavior. *Psychological Inquiry*, *11*(4), 227–268.

Diener, E., Emmons, R. A., Larsen, R. J., & Griffin, S. (1985). The satisfaction with life scale. *Journal of Personality Assessment*, *49*(1), 71–75. https://doi.org/10.1207/s15327752jpa4901_13

Diener, E., Wirtz, D., Tov, W., Kim-Prieto, C., Choi, D., Oishi, S., & Biswas-Diener, R. (2010). New well-being measures: Short scales to assess flourishing and positive and negative feelings. *Social Indicators Research*, *97*(2), 143–156. https://doi.org/10.1007/s11205-009-9493-y

Dweck, C. S. (2017). From needs to goals and representations: Foundations for a unified theory of motivation, personality, and development. *Psychological Review*, *124*(6), 689.

Epley, N., & Dunning, D. (2000). Feeling "holier than thou": Are self- serving assessments produced by errors in self- or social prediction? *Journal of Personality and Social Psychology*, *79*, 861–875.

Fiske, S. T., Cuddy, A. J., & Glick, P. (2007). Universal dimensions of social cognition: Warmth and competence. *Trends in Cognitive Sciences*, *11*(2), 77–83.

Fleeson, W. (2001). Toward a structure- and process-integrated view of personality: Traits as density distributions of states. *Journal of Personality and Social Psychology*, *80*(6), 1011–1027.

Fleeson, W. (2004). Moving personality beyond the person-situation debate: The challenge and the opportunity of within-person variability. *Current Directions in Psychological Science*, *13*(2), 83–87.

Fleeson, W., Furr, R. M., Jayawickreme, E., Meindl, P., & Helzer, E. G. (2014). Character: The prospects for a personality-based perspective on morality. *Social and Personality Psychology Compass*, *8*(4), 178–191.

Fleeson, W., & Jayawickreme, E. (2015). Whole trait theory. *Journal of Research in Personality*, *56*, 82–92. https://doi.org/10.1016/j.jrp.2014.10.009

Fleeson, W., & Jayawickreme, E. (2017). Challenging Doris' attack on aggregation: Why we are not left "completely in the dark" about global virtues. *Ethical Theory and Moral Practice*, *20*(3), 519–536.

Goodwin, G. P. (2015). Moral character in person perception. *Current Directions in Psychological Science*, *24*(1), 38–44. https://doi.org/10.1177/0963721414550709

Goodwin, G. P., Piazza, J., & Rozin, P. (2014). Moral character predominates in person perception and evaluation. *Journal of Personality and Social Psychology*, *106*(1), 148–168. https://doi.org/10.1037/a0034726

Graham, J., Haidt, J., & Nosek, B. A. (2009). Liberals and conservatives rely on different sets of moral foundations. *Journal of Personality and Social Psychology*, *96*(5), 1029–1066.

Guillen-Royo, M., & Kasser, T. (2015). Personal goals, socio-economic context and happiness: Studying a diverse sample in Peru. *Journal of Happiness Studies*, *16*(2), 405–425. https://doi.org/10.1007/s10902-014-9515-6

Hartley, A. G., Furr, R. M., Helzer, E. G., Jayawickreme, E., Velasquez, K. R., & Fleeson, W. (2016). Morality's centrality to liking, respecting, and understanding others. *Social Psychological and Personality Science*, *7*(7), 648–657. https://doi.org/10.1177/1948550616655359

Helzer, E. G., Furr, R. M., Hawkins, A., Barranti, M., Blackie, L. E. R., & Fleeson, W. (2014). Agreement on the perception of moral character. *Personality and Social Psychology Bulletin*, *40*(12), 1698–1710. https://doi.org/10.1177/0146167214554957

Hepach, R., Vaish, A., & Tomasello, M. (2017a). Children's intrinsic motivation to provide help themselves after accidentally harming others. *Child Development*, *88*(4), 1251–1264. https://doi.org/10.1111/cdev.12646

Hepach, R., Vaish, A., & Tomasello, M. (2017b). The fulfillment of others' needs elevates children's body posture. *Developmental Psychology*, *53*(1), 100–113. https://doi.org/10.1037/dev0000173

James, W. (1878). Remarks on spencer's definition of mind as correspondence. *Journal of Speculative Philosophy*, *12*, 1–18.

Jayawickreme, E., & Fleeson, W. (2017). Does whole trait theory work for the virtues? In W. S. Sinnott-Armstrong & C. B. Miller (Eds.), *Moral psychology, volume v: Virtue and happiness* (pp. 75–103). Cambridge: MIT Press.

Jayawickreme, E., Meindl, P., Helzer, E. G., Furr, R. M., & Fleeson, W. (2014). Virtuous states and virtuous traits: How the empirical evidence regarding the existence of broad traits saves virtue ethics from the situationist challenge. *Theory and Research in Education*, *12*, 1–26.

Jayawickreme, E., Zachry, C. E., & Fleeson, W. (2019). Whole trait theory: An integrative approach to examining personality structure and process. *Personality and Individual Differences*, *136*, 2–11.

Kanngiesser, P., Köymen, B., & Tomasello, M. (2017). Young children mostly keep, and expect others to keep, their promises. *Journal of Experimental Child Psychology*, *159*, 140–158. https://doi.org/10.1016/j.jecp.2017.02.004

Kristjánsson, K. (2020). *Flourishing as the aim of education*. London: Routledge. https://doi.org/10.4324/9780429464898

Mammen, M., Köymen, B., & Tomasello, M. (2018). The reasons young children give to peers when explaining their judgments of moral and conventional rules. *Developmental Psychology*, *54*(2), 254–262. https://doi.org/10.1037/dev0000424

Martela, F., & Ryan, R. M. (2016). The benefits of benevolence: Basic psychological needs, beneficence, and the enhancement of well-being. *Journal of Personality*, *84*, 750–764. https://doi.org/10.1111/jopy.12215.

Martela, F., & Ryan, R. M. (2020). Distinguishing between basic psychological needs and basic wellness enhancers: The case of beneficence as a candidate psychological need. *Motivation and Emotion*, *44*, 116–133.

Mazar, N., Amir, O., & Ariely, D. (2008). The dishonesty of honest people: A theory of self-concept maintenance. *Journal of Marketing Research*, *45*, 633–644. https://doi.org/10.1509/jmkr.45.6.633

Meindl, P., Jayawickreme, E., Furr, R. M., & Fleeson, W. (2015). A foundation beam for studying morality from a personological point of view: Are individual differences in moral behaviours and thoughts consistent? *Journal of Research in Personality*, *59*, 81–92.

Ntoumanis, N., & Standage, M. (2009). Morality in sport: A self-determination theory perspective. *Journal of Applied Sport Psychology*, *21*(4), 365–380. https://doi.org/10.1080/10413200903036040

Pierrakos, O., Prentice, M., Silverglate, C., Lamb, M., Demaske, A., & Smout, R. (2019). Reimagining engineering ethics: From ethics education to character education. In *2019 IEEE Frontiers in Education Conference (FIE)* (pp. 1–9). IEEE.

Pleasant, A., & Barclay, P. (2018). Why hate the good guy? Antisocial punishment of high cooperators is greater when people compete to be chosen. *Psychological Science*, *29*(6), 868–876. https://doi.org/10.1177/0956797617752642

Prentice, M., Jayawickreme, E., & Fleeson, W. (2019a). Integrating whole trait theory and self-determination theory. *Journal of Personality*, *87*(1), 56–69.

Prentice, M., Jayawickreme, E., & Fleeson, W. (2020). An experience sampling study of the momentary dynamics of moral, autonomous, competent, and related need satisfaction, moral exactments and psychological thriving. *Motivation and Emotion*, *44*, 244–256. https://doi.org/10.1007/s11031-020-09829-3

Prentice, M., Jayawickreme, E., Hawkins, A., Hartley, A., Furr, R. M., & Fleeson, W. (2019b). Morality as a basic psychological need. *Social Psychological and Personality Science*, *10*(4), 449–460.

Rothschild, Z. K., & Keefer, L. A. (2017). A cleansing fire: Moral outrage alleviates guilt and buffers threats to one's moral identity. *Motivation and Emotion*, *41*(2), 209–229. https://doi.org/10.1007/s11031-017-9601-2

Ryan, R. M. (1995). Psychological needs and the facilitation of integrative processes. *Journal of Personality*, *63*, 397–427.

Ryan, R. M., Curren, R. R., & Deci, E. L. (2013). What humans need: Flourishing in Aristotelian philosophy and self-determination theory. In A. S. Waterman (Ed.), *The best within us: Positive psychology perspectives on eudaimonia* (pp. 57–75). American Psychological Association. https://doi.org/10.1037/14092-004

Ryan, R. M., & Deci, E. L. (2017). Basic psychological needs theory: Satisfaction and frustration of autonomy, competence, and relatedness in relation to psychological wellness and full functioning. In R. M. Ryan & E. L. Deci (Eds.), *Self-determination theory: Basic psychological needs in motivation, development, and wellness* (pp. 239–270). New York: Guilford Publishing.

Sheldon, K. M. (2011). Integrating behavioural-motive and experiential-requirement perspectives on psychological needs: A two process model. *Psychological Review*, *118*(4), 552–569.

Sheldon, K. M., Elliot, A. J., Kim, Y., & Kasser, T. (2001). What is satisfying about satisfying events? Testing 10 candidate psychological needs. *Journal of Personality and Social Psychology*, *80*(2), 325–339. https://doi.org/10.1037/0022-3514.80.2.325

Sheldon, K. M., Sommet, N., Corcoran, M., & Elliot, A. J. (2018). Feeling interpersonally controlled while pursuing materialistic goals: A problematic combination for moral behaviour. *Personality and Social Psychology Bulletin*, *44*(9), 1330–1349. https://doi.org/10.1177/0146167218766863

Snow, N. E., Wright, J. C., & Warren, M. T. (2019). Virtue measurement: Theory and application. *Ethical Theory and Moral Practice*. https://doi.org/10.1007/s10677-019-10050-6

Steele, C. M. (1988). The psychology of self-affirmation: Sustaining the integrity of the self. In L. Berkowitz (Ed.), *Advances in experimental social psychology* (Vol. 21, pp. 261–302). Academic Press. https://doi.org/10.1016/S0065-2601(08)60229-4

Strohminger, N., & Nichols, S. (2014). The essential moral self. *Cognition*, *131*(1), 159–171. https://doi.org/10.1016/j.cognition.2013.12.005

Su, R., Tay, L., & Diener, E. (2014). The development and validation of the Comprehensive Inventory of Thriving (CIT) and the Brief Inventory of Thriving (BIT). *Applied Psychology: Health and Well-Being*, *6*(3), 251–279.

Talevich, J. R., Read, S. J., Walsh, D. A., Iyer, R., & Chopra, G. (2017). Toward a comprehensive taxonomy of human motives. *PloS One*, *12*(2), e0172279.

Tetlock, P. E., Kristel, O. V., Elson, S. B., Green, M. C., & Lerner, J. S. (2000). The psychology of the unthinkable: Taboo trade-offs, forbidden base rates, and heretical counterfactuals. *Journal of Personality and Social Psychology*, *78*(5), 853–870. https://doi.org/10.1037/0022-3514.78.5.853

Tomasello, M. (2009). *Why we cooperate* (1st ed.). Cambridge: MIT Press.

Tomasello, M., & Vaish, A. (2013). Origins of human cooperation and morality. *Annual Review of Psychology*, *64*(1), 231–255. https://doi.org/10.1146/annurev-psych-113011-143812

Vansteenkiste, M., Ryan, R. M., & Soenens, B. (2020). Basic psychological need theory: Advancements, critical themes, and future directions. *Motivation and Emotion*, *44*, 1–31.

Waal, F. de. (2009). *Primates and philosophers: How morality evolved*. Princeton: Princeton University Press.

Weinstein, N., & Ryan, R. M. (2010). When helping helps: Autonomous motivation for prosocial behavior and its influence on well-being for the helper and recipient. *Journal of Personality and Social Psychology*, *98*(2), 222–244. https://doi.org/10.1037/a0016984

Zeinoun, P., Daouk-Oyry, L., Choueiri, L., & van de Vijver, F. J. R. (2017). Arab-Levantine personality structure: A psycholexical study of modern standard Arabic in Lebanon, Syria, Jordan, and the West Bank. *Journal of Personality*. https://doi.org/10.1111/jopy.12324

8 Virtue and the specificity principle

An integrative approach to understanding character

Elise D. Murray and Jonathan M. Tirrell

Introduction

Aristotle noted that virtues are well-defined and categorised, as well as dynamic, requiring that specific virtues to be applied in specific contexts for specific individuals – where wisdom regarding such practical action is termed *phronesis*. Bridging this philosophical concept with the developmental sciences, we frame virtue within a relational developmental systems (RDS) metatheory involving mutually influential relations between persons and contexts. This chapter offers a unique perspective in that it does not see the universal-versus-local virtues debate as a dichotomy, but rather as one that can be resolved through integration and reframing as to *which aspects* of virtue are understood as universal and local. Although, on the one hand, we describe the principles of RDS metatheory as providing a universal frame for understanding and defining virtue, we, on the other hand, use the concepts of phronesis, the specificity principle, and coherence to illuminate the context-specific applications of virtue that demand attention to locality in the enactment of virtue. We illustrate this approach by describing two virtues in particular – forgiveness and humility – and conclude by discussing implications for future research and practice.

Virtue and the specificity principle: an integrative approach to understanding character

Since at least the time of Socrates, Plato, and Aristotle, philosophers have sought to identify and describe the intricate constellations of thoughts, feelings, and actions that promote good character and thriving societies. Indeed, debates have persisted for millennia regarding the universality and/or plurality of virtues, and about the systems of morality that deem them so. Today, an increasingly global and technologically advanced civilization has afforded scientists and society vast bodies of knowledge, along with an expanding appreciation of diverse perspectives and cultural frameworks that define morality and character. As such, it would be pretentious for us to claim that we will resolve debates about the nature of virtue in this chapter.

Nonetheless, we have reason to persist in our efforts of understanding virtue and character in regard to universality and locality. In societies that are increasingly diverse and multicultural, individuals and groups are often confronted with people representing varying cultural backgrounds and espousing differing moral codes (e.g., Blum, 2007). If reconciliation among, or compromise of, divergent moral priorities cannot be achieved, debates and misunderstandings risk devolving into conflict, and civility risks devolving into violence and tribalism (e.g., see Mayo, 2002; Moon, 1993). Despite the philosophical viability of moral relativism, the practical application of such a flexible lens to morality is problematic in the social and behavioral sciences, which rely heavily on generalisable patterns in order to benefit populations. That is, values specific to hegemonic groups risk eclipsing the realisation of social justice for all.

These risks therefore present us with a problem. What can be described as *universal* regarding virtue and character? Can any such understanding be applied to promoting thriving for *all*? If so, what can be learned from the virtually infinite array of context- and culture-specific virtues and manifestations of character?

In the present chapter, we situate virtue and character within contemporary models of the social and behavioural sciences framed by a relational developmental systems (RDS) metatheory involving mutually influential relations between persons and contexts (Overton, 2015). Although, on the one hand, we describe the principles of RDS metatheory as providing a universal frame for understanding and defining virtue, we, on the other hand, use the concepts of phronesis, the specificity principle, and coherence to illuminate the context-specific applications of virtue that demand attention to locality in the enactment of virtue. We illustrate this approach by describing two virtues in particular – forgiveness and humility – and conclude by discussing implications for future research and practice.

Regarding the main question animating this volume, about virtues as local or universal, the following argument offers a unique perspective in that it does not see the debate as a dichotomy, but rather as one that can be resolved through integration and reframing as to *which aspects* of virtue are understood as universal and local. We propose that the developmental relations between persons and contexts set up the universal frame as to what constitutes a virtue and, in turn, that the myriad individual and contextual variables involved in those relations demand attention to specificity, which we frame as the locality involved in the enactment of virtue.

Relational developmental systems metatheory and character science

Derived from a process-relational paradigm, RDS metatheory views human development as involving universal functions of an inherently active, living, open, self-constructing (autopoietic), self-organising, self-regulating (agentic), and integrated/holistic system that is complex and adaptive (Overton, 2015). Thus, the integration of different levels of organisation – from the biological, cellular level, to the level of social institutions (see Jablonka & Lamb, 2005; Meaney, 2010, 2014) – frames understanding of development across the life span and, as such, emphasises the concept of mutually influential relations among individuals and their contexts across ontogeny, represented as individual ⇔ context relations. These relations, or *developmental regulations*, serve as the unit of analysis in the exchanges between individuals and their contexts (Brandtstädter, 1998; Lerner, 2006). When these developmental regulations are mutually beneficial for both individual and context, they may be termed *adaptive* developmental regulations (Brandtstädter, 1998).

In contrast to RDS-based perspectives of human development, Cartesian-split, mechanistic metatheory (Overton, 2015) is a common approach used to conceptualise character, and frames psychological approaches that emphasise internal constructs and processes as the paramount correlates of character. The drawback of such existing theories is that these approaches to character research are often focused on specific components of the holistic system (e.g., the development of moral reasoning and cognition, separate from emotional and physiological development; or the development of moral character, disregarding the interplay of other domains of character such as performance, civic, and intellectual character; see Seider, 2012). In turn, RDS-based theories create a conception of a more integrated and, frankly, more interesting science that embraces much more complex dynamics and non-ergodic (i.e., non-average) change (e.g., Nesselroade & Molenaar, 2010; Rose, 2016). The common "trait approach" to character falls into this mechanistic frame (e.g., Noftle, Schnitker, & Robins, 2011), where character attributes are considered part of nature-based personality and fixed, and the emphasis is placed on the behaviours and thoughts of an individual, with the goal of detecting and tracking trait stability over time and place (Costa & McCrae, 1980; McCrae et al., 2000). Despite efforts to acknowledge and empirically explore the short-term situational dynamics of character, goals of such research is still to highlight trait stability in the longer-term (e.g., Zachry, Phan, Blackie, & Jayawickreme, 2018). Much is also extrapolated from point-in-time estimates, as opposed to longitudinal assessments of character and related individual and contextual processes (e.g., McGrath, 2015). The individual and context – as well as the processes that proceed from the individual, the context, and their coactions over time – can only be understood in their entirety in reference to one another, never as mutually exclusive entities. Therefore, both are necessary and basic components of human development (Lerner, 2018).

The developmental system is sensitive to within- and between-person differences within and across time, and thus has the potential for change, or *plasticity*, on both the individual and group levels (Lerner, 2015). Therefore, character is not a trait-like phenomenon; that is, it is not fixed (e.g., by genetics) across time and place (Nucci, 2017; Sokol, Hammond, & Berkowitz, 2010; see also Lerner & Callina, 2014; Roberts, Walton, & Viechtbauer, 2006). The malleability of human development means that conditions may be found to foster mutually adaptive relations between individuals and their contexts (Lerner, 2006). The presence of systematic variation in the observed trajectories of intraindividual change – that is, the instantiation of plasticity – allows for an optimistic approach to studying human life and the processes enabling life. Indeed, the potential for plasticity in human development underscores the idea that all processes of human life are developmental phenomena, including character.

This integrative foundation of RDS-based theories is therefore of potential use in understanding character development and the development of virtues. Instead of being understood as a trait, from an RDS-based perspective, "character . . . is a multidimensional, relatively plastic process of adaptive developmental regulations, of mutually beneficial individual ⇔ context relations" (Lerner & Callina, 2014, p. 333). Character-related theory and research from an RDS-based perspective could generate knowledge of associations between the individual and his or her specific context, and of how those associations may promote or hinder well-being and thriving on both the personal and societal levels. In addition, RDS-based theories can elucidate how these person ⇔ context relations relate to broader levels of the ecology to influence development of character within and across individuals.

In sum, from an RDS-based standpoint, human development exists in person ⇔ context relations among the multiple, fused levels of an individual's context over time. Therefore, what might be considered the *universal* aspects of any component of life are the mutually influential individual ⇔ context relations across each person's life span – a point made by biologists in the nineteenth century (e.g., Darwin, 1859), throughout the twentieth century (e.g., Lehman, 1959; Schneirla, 1957), and into the twenty-first century (e.g., Cantor, Osher, Berg, Steyer, & Rose, 2019). That is, no person, situation, or behavior is an island – there always exist contextual relations that mutually define and influence the other. Following this logic, then, a universal aspect of positive development or thriving would entail mutually *beneficial* person ⇔ context relations, described earlier as adaptive developmental regulations (Brandtstädter, 1998). In turn, a behaviour or a person may be deemed virtuous if mutually beneficial person ⇔ context relations are promoted (see Lerner & Callina, 2014).

Specificity principle: fusing phronesis and RDS metatheory

Callina and Lerner (2017) proposed that using an RDS-based approach that allows for situational flexibility based on the appropriateness of the virtue required in a specific situation would eliminate the conceptual problems of trait-oriented theories (e.g., those that disregard or do not account for systematic development or context-based sources of covariance; see

Costa & McCrae, 1980; McCrae et al., 2000; McGrath, 2015). Using RDS as a framework, instead, looks at similarities across these situations with similar individuals, to assess whether there is a consistency of how virtues are enacted in specific scenarios, and that the constellation of virtues is coherent across these situations. How, then, can this concept of universality in RDS metatheory be reconciled with the array of context- and culture-specific virtues? Indeed, elucidating what social science typifies as "universal" might seem to create a chaotic picture for systematic empirical investigation (e.g., by posing difficulties in operationalisation or criticism that we are asserting Western cultural hegemony), particularly when considering the virtually infinite array of individual and contextual variables that allow for, and support, plasticity. Nonetheless, developmental science offers, instead, an alternative approach to understanding and qualifying what is universal by providing a principle for evaluating the specific features of adaptive developmental regulations.

From an Aristotelian perspective, phronesis is a term used to describe the wisdom of practical action that informs what virtue to apply, to whom, at what time, in what ways, and in what amount. Recently, phronesis has become a tool for psychologists to frame multidimensional approaches to understanding mainly wisdom (e.g., Grossmann, 2017). In more recent attempts, and best aligned with the developmental considerations of character, Darnell, Gulliford, Kristjansson, and Paris (2019) used phronesis in order to provide a multi-component solution in moral psychology and character research, to what is known as the "gappiness problem", wherein moral action research lacks a link, or has a gap, between moral judgment and moral action. Their position is that this gap has so far only been filled with single-component solutions – that is, moral identity (e.g., Blasi, 1980) and moral emotions (e.g., Hoffman, 2000) – with echoes of the Cartesian-split, reductionist approaches of character science broadly. To address the process by which moral actions occur, Darnell et al. (2019) instead provide a phronesis-based theoretical frame, consisting of: (1) a constitutive function, or the ability to recognise aspects of one's circumstance to act accordingly; (2) an integrative function, or the ability to create a response to the aspects of the context seemingly in conflict; (3) a blueprint of the good life; and (4) emotional regulation.

The concept of positive development – that is, one that focuses on mutually beneficial relations among individuals and contexts – marries well with the concept of phronesis, such that the array of individual- and context-specific factors requires attention to specificity in how a specific virtue is appropriate at a specific time and place. Darnell et al. (2019) show the potential for phronesis to explain the process of moral action in a way that is representative of the complexity and unique nature of a specific moral action, as well as the structure of such a process that can be explained using a universal frame. That is, using a broad frame that embraces multidimensionality of the processes between moral judgment and action allows for situational specificity of an action without violating any notions of what is and is not virtuous (should the phronesis be employed appropriately).

In developmental science, such exactness according to situational factors and individual differences in phronesis is not only consistent with RDS metatheory and mutually beneficial individual ⇔ context relations, but it is also consistent with the specificity principle (Bornstein, 2017). This principle entails asking a series of *what* questions: what works (e.g., about a program), for what specific individuals, in what specific contexts, at what specific times (historically and developmentally), for what specific domains of development, and in what specific ways. This principle is a way to apply RDS metatheory to mid-range theories, as the "specifics" asked in the principle reflect the practical levels of questions that pinpoint all levels of an individual's ecology as well as appropriateness and effectiveness of a phenomenon for a specific person in a specific context. In regard to acting virtuously, as Darnell et al. (2019) note, phronesis may be regarded as the executive functioning tool useful for discerning what behaviours or responses would be appropriate given a specific situation, and therefore determines what amount of what virtue is best for that circumstance. Similarly, phronesis, we argue, can be used to understand the nature of virtues themselves, with the additional support of RDS framing and the specificity principle.

From an RDS-based frame using the specificity principle, a virtue itself is only a virtue if its application is sensitive to the mutually influential individual ⇔ context relations of a specific moral situation in such a way that it promotes mutually *beneficial* relations. Such specificity suggests that nothing is universal – more precisely, that group-based averages do not and cannot represent any one individual (Rose, 2016) – given situational contingencies and between- and within-individual differences. This sentiment may be aligned, then, with the ideas of moral relativism, that virtues are, in fact, hyper-local due to their dependence on the individual and contextual factors involved in a given situation.

How, then, does a virtue have any universal attributes, especially considering the specificity principle? Despite this highly specific, individual ⇔ context conceptualisation of virtue, the construct still requires a universal frame, a concrete definition by which comparisons of mutually beneficial individual-context relations can be derived. Therefore, we cannot entirely align our approach to character with that of moral relativists, because we frame what is perceived as relativistic as, rather, an individual-context specific manifestation of virtues that are universal. From a developmental science perspective, then, virtues do not exist as disembodied or decontextualized laws that are themselves considered universal or local; rather, virtues emerge from the *relations* between and among persons and contexts, the quality of which necessarily varies according to locality.

Nucci (2017) and Callina and Lerner (2017) would qualify this integrated movement toward an RDS-based, phronesis-focused model of virtue as one of "coherence" of character, rather than consistency of action. Indeed, to *consistently* apply a behaviour that might be considered virtuous in some contexts, without regard to the specific needs of the situation, lacks a logical *coherence* that might represent an optimal way of being virtuous in the world. This aspect of coherence is closely aligned with the integrative function of Darnell et al.'s (2019) phronesis-based model for moral action. However, we wish to apply the concept more broadly; that is, seemingly conflicting series of moral actions are integrated into a coherent moral profile, of sorts. The strengths of the flexible model presented by Darnell et al. (2019) are supportive of developmental methods and paradigms, but their moral action theory does not focus on the recursive process that is required to more strongly develop a particular virtue. Rather, their conceptualisation focuses on the sensitivity within each situation for a virtue to be elicited in a sequence and not the *development* of that process, and therefore could still succumb to trait conceptualisations without a strong developmental backdrop. Therefore, there is still an overwhelming tendency in character science to prioritise consistency or stability of character, as compared to considering coherence across situational variation.

This notion of coherence, therefore, encompasses and allows for: (1) a universality of virtue, in that broad conceptualisations of a specific virtue involve adaptive developmental regulations (i.e., those that promote mutually beneficial individual ⇔ context relations); and, (2) specificity, or locality, of the dynamics and appropriateness of the virtues across and within individuals. Phronesis, then, is the executive functioning tool for navigating such specificity in deploying virtue (Darnell et al., 2019; Schwartz & Sharpe, 2006).

If we are to usefully and empirically explore these concepts, character researchers must integrate philosophical and scientific assessments of virtue. That is, trait-based theories, and point-in-time estimates, of character must be discussed and aligned with the RDS-based and specificity principle-based theories of virtue. This theoretical work must involve models that are sensitive to this understanding of the developmental components of phronesis, as explained by Darnell et al. (2019) regarding the process of moral action.

Applying the specificity principle to virtues: two examples

We use two examples of moral virtues – forgiveness and humility – to illuminate this theoretical model of character for integrating the universal and local aspects of virtue. According to the RDS-based model focused on phronesis and coherence, the *universal* aspect of virtue is that it promotes mutually beneficial person ⇔ context relations, or adaptive developmental regulations. Given the vast array of person- and context-specific variables across time and place, the realisation of such universality – that is, in order to understand virtues as promoting mutually beneficial relations – means that *locality* in regard to manifestations and applications of virtue must necessarily vary according to the specificity principle. Phronesis, then, being in effect an executive functioning tool (e.g., cognitive flexibility and behaviour inhibition) for navigating specificity, enables discerning the practical action to take that may promote personal and societal well-being which, in turn, may be termed virtuous action.

Forgiveness

Forgiveness involves "a willingness to abandon one's right to resentment, negative judgment, and indifferent behavior toward one who unjustly injured us, while fostering the undeserved qualities of compassion, generosity, and even love toward him or her" (Enright, Freedman, & Rique, 1998, p. 47). Its role as a virtue, however, remains controversial and, from a developmental science perspective, understudied empirically. Although research supports health benefits for the forgiver in managing and letting go of negative emotions (see Toussaint, Worthington, & Williams, 2015), much less is known about whether forgiveness promotes *mutually* beneficial relations in human development – that is, do forgiveness responses to injustice benefit the forgiven as well as the forgiver, the persons, and the context?

Challenges to forgiveness are not new. Nietzsche (1996) regarded forgiveness as a weakness – a view also held by some clinicians and scholars who challenge the value of forgiveness as a virtue. For instance, Exline, Worthington, Hill, and McCullough (2003) describe how some motives to forgive may reflect vice behaviors, including fear of confrontation, avoidance of anger, or prioritising personal comfort over justice. Others conflate forgiveness with pardoning or justifying bad behavior, and being "too soft on justice" (Worthington, 2005, p. 21). Given such controversy, how could forgiveness be considered a character virtue, and how could such a "virtue" be considered universal?

We argue that such views of forgiveness as a vice behaviour may reflect instances of rigid consistency, lacking the coherence allowed by phronesis, in which the specificity principle is applied with the aim of promoting mutually beneficial relations between person and context. We could imagine instances when such rigid consistency of forgiveness responses across contexts might yield outcomes that are less than mutually beneficial. For instance, consider a victim who forgives a thief out of fear of confrontation or retaliation: the victim may initially feel relief from the negative emotions by transforming them into positive ones; yet the avoidance-motivated response might prevent the thief from being prosecuted, which, in turn, might lead the thief to steal again, putting the community at risk for repeat offenses. Such an instance may not be termed "true" forgiveness, let alone virtuous.

Similarly, another example might involve one who takes a deontological approach to forgiveness, always forgiving interpersonal grievances because forgiveness is deemed the "right thing to do". The forgiver may risk neglecting his or her own negative experiences, missing opportunities to process and learn from the injustices, and the offender may not be responded to in a way in which positive change can occur; again, risking repeat offenses. In these instances, there is a lack of promoting *mutually* beneficial outcomes for forgiver and forgiven (and the community context) and, as such, a lack of justification, from an RDS-based perspective, for counting forgiveness as a virtue.

However, when the proposed universal aspects of virtue are considered – that virtue involves promoting mutually beneficial relations between persons and contexts; and that phronesis, or applying the specificity principle, is used to discern the response that promotes mutually beneficial relations – a coherent view of forgiveness as a virtue may emerge. For instance, the previously noted victims may use forgiveness, first to recognise, reflect upon, and process their own negative responses to injustice (the emotional and cognitive responses of forgiveness), and then they may act in such a way (the behavioural response) that promotes justice while seeking to redress the humanity of the offender.

One classic example of such a response may be Pope John Paul II's forgiveness of his attempted assassinator, while the transgressor was still serving time in prison. There was no condoning or pardoning the offence; yet forgiveness allowed the victim to foster positive responses, the offender to be seen as a person with value and dignity, and, perhaps, the community to grow toward restorative justice. Indeed, as Pope John Paul II proclaimed, "The command to forgive does not precede the objective demands of justice, but justice in the correct sense of the word is actually the ultimate aim of forgiveness" (Encyclical no. XIV, quoted in Bronkhorst, 1995, p. 41). Forgiveness, itself, does not need to be consistently ever-present for one to be considered virtuous; instead, the goal of promoting thriving for all involved in the situation, and using phronesis to determine the appropriate response to the specific instance of injustice, guides the forgiving response toward virtue and, as Pope John Paul II described, justice.

More work is indeed needed to further explore such a model of forgiveness as a virtue. Nonetheless, evidence does exist suggesting that forgiveness responses to injustice may promote mutually beneficial outcomes for the forgiver, the forgiven, and the community. For instance, in regard to school bullying, forgiveness-related interventions have been associated with reducing bullying behaviours and promoting harmonious communities (e.g., Ahmed & Braithwaite, 2006; Hui, Tsang, & Law, 2011). In regard to criminal behaviour, restorative justice approaches promoting forgiveness have been associated with decreased recidivism (e.g., Sherman, Harris, & Erbes, 2015) and promoting positive transformational change among offenders (e.g., Adler & Mir, 2012; Brown, 2011; Straub, 2013).

Certainly, in each instance of school bullying and criminal behaviour, a rigid and consistent, decontextualised forgiveness response would not produce such positive outcomes, which contributes to our argument that we do not consider virtues themselves to be universal. Rather, we can expect that a level of phronesis was applied to understand what specific response, to what specific individual, in what specific amount, at what specific time, and in what specific way, promoted such mutually beneficial outcomes that would qualify a virtue. As such, the locality of the relations in which the forgiving response was enacted determines whether the response was virtuous. In other words, while the person ⇔ context relations involved in human development are universal, the manifestation of a virtue is local, qualified by the promotion of mutually beneficial relations.

Humility

Humility also garners controversy akin to that of forgiveness. At its most basic level, similar to other virtue constructs (Card, 2017), there still exists debate about the "right" way to conceptualise humility (e.g., Nadelhoffer, Wright, Echols, Perini, & Venezia, 2016). A lack of definitional consensus leads to problems of consistent measurement, particularly when different paradigms drive a particular empirical question related to that definition of humility. Aristotle's theory of virtuous means denotes that pride, not humility, is actually the virtue to strive toward, a balance between arrogance and servility (or humility). This conception of proper pride is that there is an accurate individual perception of strengths. Many faith traditions took the same coin and addressed the opposite side, in that, with accurate perceptions of strengths comes the accurate perceptions of limitations and, thus, humility. But this view of humility tended toward servility, that of the Christian tradition, which raised an issue with those attempting to reconcile Aristotle's non-overestimation in operationalising and measuring this construct in the psychological sciences. Despite newer conceptualisations of humility, this movement has not led to consensus. In their chapter on the various philosophical definitions of humility, Roberts and Cleveland (2016) list some of the competing definitions of humility, including humility as chronic low self-esteem, ignorance of one's ability, limitations-owning, and inattention to one's ability. For the purposes of using an RDS-based definition, we define humility in this chapter from the relational humility lens, which involves an accurate view of the self with modest self-presentation and an orientation toward others (Van Tongeren & Myers, 2016).

In considering the literatures of philosophy and psychology, this perspective of humility is both satisfying and frustrating; it holistically accounts for an accurate view of the self, but creates challenges for operationalisation and measurement. Accuracy requires an appropriate understanding of both someone's inabilities as well as strengths, which means that definitions could argue that confidence is somehow part of humility and therefore a requisite component of the construct (as compared to being a separate but necessary component of the process of enacting humility; see Church, 2017 in relation to intellectual humility). Furthermore, as an accurate view of the self is person- and context-specific, operationalisation of a general, consistent construct of humility is difficult.

One positive component involved in scholarship about humility is that its research has lent itself quite easily to the framework of virtues as represented by mutually beneficial relations. That is, humility research acknowledges an orientation toward the other in the process of exhibiting the virtue. Given its importance in relationship development and maintenance, humility research has managed to extend past typical psychological research and expand into management studies and other applied fields. For example, leader humility and the effects on organisations has been a major focus of work in management schools (e.g., Owens & Hekman, 2016) and has led to a movement in larger corporations hiring for capabilities such as *intellectual* humility (an attribute Google seeks out in future employees; see Baer, 2014). Research has also emerged about the importance of humility in marriages (e.g., Wang, Edwards, & Hill, 2017) and, unsurprisingly, potential marital improvements if both partners are humble (Van Tongeren et al., 2017). The controversy in humility research emerges in the definitions of humility and how it is pieced together with the concept of pride.

Issues in conceptualising the virtue and vice components of humility fall away when coherence and specificity are applied to understanding the construct, with an emphasis on the mutually beneficial relations between person and context. As noted previously, empirical investigations of humility emphasise such mutually beneficial relations. Humility can be vital to knowledge acquisition and improving contexts through such knowledge, and, as mentioned previously, relationship-building and maintenance in various private and public spheres of one's life (e.g., Owens & Hekman, 2016; Wang et al., 2017). Humility is additionally beneficial for developing other-person-directed social-emotional competencies, such as empathy and other forms of perspective-taking (e.g., Davis et al., 2011), and even benevolence (Wright et al., 2017).

Nevertheless, there are circumstances when humility might not serve individuals acting deferentially to others but, instead, to enact a sense of confidence and authority. An example from on-going work at the US Military Academy on character is that, in times of warfare and on-the-ground decision-making, the most prudent course of action is not for the leader of a corps of soldiers to ask his or her subordinates for suggestions on the next step, but rather to take charge and ensure the safety of the individuals in his or her command. Doing otherwise could result in immense harm, and therefore humility is not the appropriate virtue to enact; that is, in this instance, humility would not result in mutually beneficial relations between person and context. Whereas the leader may gain some insight and knowledge from those who have been enlisted soldiers with more field experience (person-benefitting), most likely great danger will befall the corps if the leader delays action and hence invites harm.

In developing a narrative of coherence and specificity, the importance of integrating phronesis into empirical investigations and theoretical approaches to humility is again brought to the fore. Similar to the argument that forgiveness, if elicited consistently across all circumstances, would at times be construed as a vice, humility is subject to the same balancing act with other virtues. Applying specificity and phronesis prompts such questions as: (1) is this virtue necessary at all in this circumstance?; and, if so, (2) what amount of this virtue is appropriate, and in what fashion, that there will be benefit to both me and the circumstance overall? In sum, these questions give a coherent, dynamic portrait of virtues wherein there is a constant process of fore-fronting and back-burnering virtues based on person, place, and time. This individual ⇔ context process allows maintenance of an integrative perspective about the universal, mutually beneficial relations that result in virtues exhibited in unique and context-specific ways.

These two examples are not exhaustive by any means, but they exemplify how vital it is to find ways to connect empirical work and applied anecdotes to the concept of coherent character, sensitive to RDS metatheory, the specificity principle, and phronesis. Similar to Darnell et al. (2019), we believe in elevating the importance of phronesis, the specificity principle, and coherence in exhibiting virtue, as compared to perspectives on consistency, both within and across individuals. This perspective on virtue does not, then, take one side or the other of the debate on the locality or universality of virtue. Rather, we suggest that there is theoretical support and evidence that virtues are universal constructs that require mutually beneficial individual-context relations, necessitating local manifestations that are sensitive to those person-specific processes.

Moving the needle on virtue in developmental science

By merging the RDS-based concept of mutually beneficial individual ⇔ context relations in human development with the specificity principle, and the specificity principle and coherence – together reflective of phronesis – the result is a model by which to both recognise and exhibit virtue in a way that is not constrained by either split trait or situational approaches. Instead, what is possible is a conception wherein universal attributes that are, in every circumstance, exhibited as a result of the unique combination of individual, context, and the relations between these components of the relational developmental system. The relations involve individual ⇔ context relations that are coherent and sensitive to change. Utilising such an approach to classifying and interpreting virtues could also assist in the inclusion/exclusion issue in virtue theory and research; a more comprehensive approach may then lend itself to more psychometrically robust measures of virtues, a pervasive issue in the social and behavioural sciences (e.g., Card, 2017).

With such an eye for the dynamic nature of the individual and his or her context, we can better understand virtue as it manifests across and within individuals, and more appropriately design contexts that provide positive spaces for furthering the development of aspects of human flourishing, such as character virtues. We can use this developmental, integrated approach to focus on measurements and analyses that address the unity and dynamics of the virtues and not just the developmental trajectories and idiosyncrasies of singular virtues. In addition, as Darnell et al. (2019) note, is not only how phronesis can be useful to character and developmental science, but also how developmental science can be applied to phronesis and the processes that affect how virtues develop. How do phronesis and coherence develop, and how are they situated in the psychological sciences so that we can use them to then understand the processes in which they most affect? Addressing the interrelations between virtues and the processes which affect virtues fully realises the tenets of RDS metatheory and the integrated nature of all levels of one's ecology. With such techniques, we can finally take the philosophical theories, armed with social science, and empirically explore what has been hypothesised about in the past.

Acknowledgements

Author correspondence should be addressed to Elise Murray, Program for Leadership and Character, Starling Hall, PO Box 7392, Wake Forest University, Winston-Salem, or dykhuiem@wfu.edu. We are grateful to Dr. Richard Lerner for his extensive feedback on this chapter.

References

Adler, J., & Mir, M. (2012). *Evaluation of the forgiveness project within prisons*. London: Middlesex University Research Repository.

Ahmed, E., & Braithwaite, V. (2006). Forgiveness, reconciliation, and shame: Three key variables in reducing school bullying. *Journal of Social Issues*, *62*(2), 347–370.

Baer, D. (2014). *Why Google wants new hires who are humble and argue*. Retrieved from www.fastcompany.com/3027116/why-google-wants-new-hires-who-are-hum ble and-argue

Blasi, A. (1980). Bridging moral cognition and moral action: A critical review of the literature. *Psychological Bulletin*, *88*(1), 1–45.

Blum, L. (2007). Race, national ideals, and civic virtue. *Social Theory and Practice: An International and Interdisciplinary Journal of Social Philosophy*, *33*(4), 533–556.

Bornstein, M. H. (2017). The specificity principle in acculturation science. *Perspectives in Psychological Science*, *12*, 3–45.

Brandtstädter, J. (1998). Action perspectives on human development. In W. Damon & R. M. Lerner (Eds.), *Handbook of child psychology: Theoretical models of human development* (Vol. 1, 5th ed., pp. 807–863). New York: Wiley.

Bronkhorst, D. (1995). *Truth and reconciliation: Obstacles and opportunities for human rights*. Amsterdam: Amnesty International Dutch Section.

Brown, T. (2011). *Interventions: Initial briefing on RESTORE – The forgiveness project prison program at HMP & YOI Parc*. S. Wales: HMP Parc Bridgend.

Callina, K. S., & Lerner, R. M. (2017). On the importance of coherence in the study of character development. *Journal of Character Education*, *13*(1), 17–25.

Cantor, P., Osher, D., Berg, J., Steyer, L., & Rose, T. (2019). Malleability, plasticity, and individuality: How children learn and develop in context. *Applied Developmental Science*, *23*(4), 307–337.

Card, N. A. (2017). Methodological issues in measuring the development of character. *Journal of Character Education*, *13*(2), 30–47.

Church, I. M. (2017). The limitations of the limitations-owning account of intellectual humility. *Philosophia*, *45*(3), 1077–1084.

Costa, P. T., Jr., & McCrae, R. R. (1980). Still stable after all these years: Personality as a key to some issues in adulthood and old age. *Life-Span Development and Behavior*, *3*, 65–102.

Darnell, C., Gulliford, L., Kristjansson, K., & Paris, P. (2019). Phronesis and the knowledge- action gap in moral psychology and moral education: A new synthesis? *Human Development*, *62*, 1–29.

Darwin, C. (1859). *The origin of species; and, the descent of man*. Modern Library.

Davis, D. E., Hook, J. N., Worthington, E. L., Jr., Van Tongeren, D. R., Gartner, A. L., Jennings, D. J., & Emmons, R. A. (2011). Relational humility: Conceptualizing and measuring humility as a personality judgment. *Journal of Personality Assessment*, *93*(3), 225–234.

Enright, R. D., Freedman, S. R., & Rique, J. (1998). The psychology of interpersonal forgiveness. In R. D. Enright & J. North (Eds.), *Exploring forgiveness* (pp. 46–62). Madison, WI: The University of Wisconsin Press.

Exline, J. J., Worthington, E. L., Jr., Hill, P., & McCullough, M. E. (2003). Forgiveness and justice: A research agenda for social and personality psychology. *Personality and Social Psychology Review*, *7*(4), 337–348.

Grossmann, I. (2017). Wisdom in context. *Perspectives on Psychological Science*, *12*(2), 233–257.

Hoffman, M. L. (2000). *Empathy and moral development: Implications for caring and justice*. Cambridge: Cambridge University Press.

Hui, E. K. P., Tsang, S. K. M., & Law, B. C. M. (2011). Combating school bullying through developmental guidance for positive youth development and promoting harmonious school culture. *The Scientific World Journal*, *11*, 2266–2277.

Jablonka, E., & Lamb, M. (2005). *Evolution in four dimensions: Genetic, epigenetic, behavioral, and symbolic variation in the history of life*. Cambridge: MIT Press.

Lehman, H. C. (1959). *Age and achievement*. Princeton: Princeton University Press.

Lerner, R. M. (2006). Developmental science, developmental systems, and contemporary theories of human development. In R. M. Lerner (Ed.), *Handbook of child psychology, volume 1: Theoretical models of human development* (6th ed., pp. 1–17). Editors-in-chief: W. Damon & R. M. Lerner. New York: Wiley.

Lerner, R. M. (2015). Preface. In W. F. Overton & P. C. Molenaar (Eds.), *Theory and method volume 1 of the handbook of child psychology and developmental science* (7th ed., pp. xv–xxi). Hoboken, NJ: Wiley.

Lerner, R. M. (2018). Character development among youth: Linking lives in time and place. *International Journal of Behavioral Development*, *42*(2), 267–277.

Lerner, R. M., & Callina, K. S. (2014). The study of character development: Towards tests of a relational developmental systems model. *Human Development*, *57*, 322–346.

Mayo, C. (2002). The binds that tie: Civility and social difference. *Education Theory*, *52*(2), 169–186.

McCrae, R. R., Costa, P. T., Jr., Ostendorf, F., Angleitner, A., Hrebickova, M., Avia, M. D., . . . Smith, P. B. (2000). Nature over nurture temperament, personality, and life span development. *Journal of Personality and Social Psychology*, *78*, 173–186.

McGrath, R. E. (2015). Character strengths in 75 nations: An update. *The Journal of Positive Psychology*, *10*(1), 41–52.

Meaney, M. (2010). Epigenetics and the biological definition of gene x environment interactions. *Child Development*, *81*, 41–79.

Meaney, M. (2014). Epigenetics offer hope for disadvantaged children. *Child & Family Blog*. Retrieved from childandfamilyblog.com/epigenetics-offer-hope-disadvantaged-children

Moon, D. J. (1993). *Constructing community: Moral pluralism and tragic conflicts*. Princeton: Princeton University Press.

Nadelhoffer, T., Wright, J. C., Echols, M., Perini, T., & Venezia, K. (2016). Some varieties of humility worth wanting. *Journal of Moral Philosophy*, *14*, 168–200.

Nesselroade, J. R., & Molenaar, P. C. M. (2010). Emphasizing intraindividual variability in the study of development over the lifespan. In W. F. Overton (Ed.), *Handbook of life-span development, volume 1: Cognition, biology, and methods* (pp. 30–54). Editor-in-chief: R. M. Lerner. Hoboken, NJ: Wiley.

Nietzsche, F. W. (1996). *The genealogy of morals* (D. Smith, Trans.). Oxford: Oxford University Press. (Original work published 1887).

Noftle, E. E., Schnitker, S. A., & Robins, R. W. (2011). Character and personality: Connections between positive psychology and personality psychology. In K. M. Sheldon, T. B. Kashdan, & M. F. Steger (Eds.), *Designing positive psychology: Taking stock and moving forward* (pp. 207–227). New York: Oxford University Press.

Nucci, L. P. (2017). Character: A multi-faceted developmental system. *Journal of Character Education*, *13*(1), 1–16.

Overton, W. F. (2015). Process and relational developmental systems. In W. F. Overton & P. C. Molenaar (Eds.), *Theory and method, volume 1 of the handbook of child psychology and developmental science* (7th ed., pp. 9–62). Editor-in-Chief: R. M. Lerner. Hoboken, NJ: Wiley.

Owens, B. P., & Hekman, D. R. (2016). How does leader humility influence team performance? Exploring the mechanisms of contagion and collective promotion focus. *Academy of Management Journal*, *59*(3), 1088–1111.

Roberts, B. W., Walton, K. E., & Viechtbauer, W. (2006). Patterns of mean-level change in personality traits across the life course: A meta-analysis of longitudinal studies. *Psychological Bulletin*, *132*, 1–25.

Roberts, R. C., & Cleveland, W. S. (2016). Humility from a philosophical point of view. In E. L. Worthington, Jr., D. E. Davis, & J. N. Hook (Eds.), *Handbook of humility: Theory, research, and applications* (pp. 33–46). New York: Routledge.

Rose, T. (2016). *The end of average: How we succeed in a world that values sameness*. New York: HarperCollins Publishers.

Schneirla, T. (1957). The concept of development in comparative psychology. In D. B. Harris (Ed.), *The concept of development* (pp. 78–108). Minneapolis: University of Minnesota Press.

Schwartz, B., & Sharpe, K. E. (2006). Practical wisdom: Aristotle meets positive psychology. *Journal of Happiness Studies*, *7*(3), 377–395.

Seider, S. (2012). *Character compass: How powerful school culture can point students toward success*. Cambridge: Harvard Education Press.

Sherman, M. D., Harris, J. I., & Erbes, C. (2015). Clinical approaches to addressing spiritual struggle in veterans with PTSD. *Professional Psychology: Research and Practice*, *46*(4), 203.

Sokol, B. W., Hammond, S. I., & Berkowitz, M. W. (2010). The developmental contours of character. In T. Lovat, R. Toomey, & N. Clement (Eds.), *International research handbook on values education and student wellbeing* (pp. 579–603). Dordrecht, Netherlands: Springer.

Straub, C. (2013). *Evaluation report: Embedding RESTORE into the fabric of YOI Ashfield – Qualitative analysis of impact and effectiveness*. London: The Forgiveness Project.

Toussaint, L. L., Worthington, E. L., Jr., & Williams, D. R. (2015). *Forgiveness and health: Scientific evidence and theories relating forgiveness to better health*. New York: Springer.

Van Tongeren, D. R., Hook, J. N., Ramos, M. J., Edwards, M., Worthington, E. L., Jr., Davis, D. E., & Cowden, R. G. (2017). The complementarity of humility hypothesis: Individual, relational, and physiological effects of mutually humble partners. *The Journal of Positive Psychology*, 1–10.

Van Tongeren, D. R., & Myers, D. G. (2016). A social psychological perspective on humility. In E. L. Worthington, D. E. Davis, & J. N. Hook (Eds.), *Handbook of humility: Theory, research, and applications* (pp. 150–164). New York: Routledge.

Wang, F., Edwards, K. J., & Hill, P. C. (2017). Humility as a relational virtue: Establishing trust, empowering repair, and building marital well-being. *Journal of Psychology and Christianity*, *36*(2), 168–179.

Worthington, E. L., Jr. (2005). *The power of forgiving*. West Conshohocken, PA: Templeton Foundation Press.

Wright, J. C., Nadelhoffer, T., Perini, T., Langville, A., Echols, M., & Venezia, K. (2017). The psychological significance of humility. *The Journal of Positive Psychology*, *12*(1), 3–12.

Zachry, C. E., Phan, L. V., Blackie, L. E., & Jayawickreme, E. (2018). Situation-based contingencies underlying wisdom-content manifestations: Examining intellectual humility in daily life. *The Journals of Gerontology: Series B*, *73*(8), 1404–1415.

9 Diversity, deviance, and virtue within imperfect moral communities

Jennifer Cole Wright

Introduction

Let us start this chapter with the following premise: *we are morally imperfect beings*. Not only do we quite often fail to do the right thing in a given situation, but we also often simply do not know what the right thing to do would be. And this makes creating, and enforcing, a system of shared norms – norms whose purpose it is to guide our social lives and protect us from injustice, harm, and oppression – tricky, to say the least.

On the one hand, we need a system of shared norms that will effectively discourage – even prevent – deviant beliefs, values, and practices from occurring. But, on the other hand, we also need to allow room for change, room for a diversity of new and different beliefs, values, and practices to be introduced and explored. Given our current state of moral imperfection, providing room for change is the only way that we can make advances in our moral understanding of ourselves – and in the system of shared norms that govern us – even if not every change we introduce and explore turns out to be actual *improvement*.

But how do we determine what is "diversity" and what is "deviance"? How do we determine when enforcing existing norms and discouraging alternative beliefs, values, and practices is necessary to keep people safe, and when it is thwarting healthy diversity and genuine moral advancement? These questions are critical for creating and maintaining healthy sociocultural normative systems (i.e., socially "normed" shared beliefs, values, attitudes, practices, commitments, rituals, etc.) that allow individuals within a community to function well (and ideally thrive).

They also bear on the question of how virtues develop, and which ones will be prioritised – i.e., those necessary for protecting existing normative systems, and those needed for corrective purposes, when normative systems become dysfunctional and change is required. This introduces space for local variation in virtue development and expression, both within and between cultures. But it also highlights a problem – the imperfection of our moral knowledge and the vulnerability of our normative systems to error and corruption. The challenge we face is having an understanding of ourselves as moral beings oriented towards "the good" that is global enough to be meaningfully shared and stable enough to be passed down to future generations, yet flexible enough to adapt and change as our cumulative experiences bring that understanding into question.

Finding answers to these questions is going to be challenging, and while I do not propose to provide them in this chapter, I will lay out some of the psychosocial mechanisms (or processes, if you would prefer) that my research suggests people actively utilise in an effort to navigate this difficult space.

Domain classification

According to the research my own lab at the College of Charleston and others have been conducting, one of the central ways people navigate this difficult space is with *domain classification* (for an overview of domain classification, and social domain theory in general, see Nucci, 1981, 1996; Smetana, 1981, 1983; Turiel, 1983, 1998). At the core of domain classification is the "social-conventional" – what I will also refer to as the "socio-cultural" – domain. This is the shared normative space we co-create – i.e., the shared system of socio-cultural norms whose purpose it is to guide our social lives so that we can co-exist together harmoniously as we engage in both collaborative and individual endeavours, making it possible for us to function as a cohesive whole.

While these norms are, at least to some degree, arbitrary (there are lots of different ways that people can peacefully co-exist in socio-cultural groups), once they have been created, members of the socio-cultural group to which they apply are expected to follow them; they are enforceable. Thus, an expectation of *conformity* exists – people will, by and large, adhere to socio-cultural norms and act in ways agreed upon and approved by their relevant groups. Failing to adhere to these norms (outside of a reasonable margin of flexibility or error) is perceived as a transgression and typically results in social punishment.

Yet, even in a tightly knit socio-cultural group, individuals require a certain amount of space to be themselves in order to thrive – the space to *not* conform to the norms of the group, but instead to have their own unique beliefs, values, and practices; to "do their own thing". This is the space of the personal domain, the domain of *personal autonomy/agency*. It is a space that is understood to be protected from unreasonable intrusion from the social group – a space that is "off limits" (at least to some degree) from the regulatory restrictions of socio-cultural norms. It is a space of ingenuity and creativity. It is also a space through which diversity is generated – where new beliefs, values, and practices are introduced to the group.

Of course, both individuals and socio-cultural groups can go astray, adopting beliefs, values, and practices that are *deviant* – i.e., that are harmful, that unjustly impose on others, that are unnecessarily inequitable, oppressive, etc. Thus, the moral domain (the final domain in the social domain theory triad), exists to protect against this. Its function is to protect socio-cultural groups from the introduction of harmful/unjust/oppressive beliefs, values, and practices. It sets limits to what socio-cultural groups – and the individuals living within them – can reasonably do (and, to a lesser degree, reasonably believe and value), or expect other members of their group to do.[1]

In summary, the personal and moral domains work together to regulate the socio-cultural domain, to protect against abuses in our shared social/cultural space. They actively balance between the need to protect individuals' space for personal autonomy from socio-cultural encroachment – to promote and protect a space for diversity – and the need to protect the whole from error and corruption, and from the introduction of deviant (harmful, unjust, and oppressive) beliefs, values, and practices.[2,3]

Over the years, my lab and others have collected a substantial amount of empirical data that supports this view (Killen & Nucci, 1995; Nichols, 2004; Nucci, 1981; Nucci & Turiel, 2000; Skitka, Bauman, & Sargis, 2005; Skitka & Mullen, 2002; Smetana, 1981, 1983; Turiel, 1983, 1998; Wainryb, Shaw, Langley, Cottam, & Lewis, 2004; Wainryb, Shaw, Laupa, & Smith, 2001; Wainryb, Shaw, & Maianu, 1998; Wright, 2012; Wright, Cullum, & Schwab, 2008). Specifically, research has found that children, adolescents, and adults actively use these domains to locate divergent beliefs, values, and practices – that is, they actively label divergent beliefs, values, and practices as "personal", "social/conventional", or "moral" – and that, even though there is disagreement (both within and between age groups) about which beliefs, values, and practices fall into the different domains, they nonetheless appear to understand the domains and use them with little evidence of confusion.[4]

What is more, their use of these domains powerfully predicts their response to those divergent beliefs, values, and practices.[5] In particular, they expressed significantly more attitudinal and behavioral tolerance for the beliefs, values, and practices they identified as personal, and significantly less attitudinal and behavioral tolerance for the ones they identified as moral, with the ones they identified as social falling in between.

More specifically, people reported being most accepting of divergent personal beliefs, values, and practices, and least accepting of moral beliefs, values, and practices, especially when they were encountered in "intimate" contexts – such as in a college roommate or best friend. They reported being most willing to *interact with* (in both intimate and non-intimate settings) and *help* (in both low and high-commitment situations) someone with divergent personal beliefs, values, and practices, and least willing to do so for someone with divergent moral beliefs, values, and practices. And while people view both social/conventional and moral divergence as problematic – since both domains involve norms that must be followed – they judged people with divergent moral beliefs, values, and practices significantly more harshly. Someone with divergent social/conventional beliefs, values, and practices was viewed as "different", as more likely to be from a different socio-cultural group, as "not like them". Someone with divergent moral beliefs, values, and practices, on the other hand, was viewed as disgusting, as being a *bad* person (Wright, 2012; Wright et al., 2008).

In a controlled experimental setting, people sat further away and turned their bodies away more from a conversational partner when they disagreed with them about a moral issue than when they disagreed with them about a personal issue. And they shared fewer raffle tickets for prizes at local stores with someone with divergent moral beliefs, values, and practices than they did with someone with divergent personal beliefs, values, and practices (Wright et al., 2008). Finally, people were most willing to prohibit, censor, shun, and punish divergent moral beliefs, values, and practices – i.e., to prevent them from entering into (and "taking root" in) their shared socio-cultural space – and most likely to condone and support divergent personal beliefs, values, and practices (Wright, 2018a), even when they didn't agree with them themselves.

In summary, we found that people utilized the personal domain for holding open – and the moral domain for shutting down – the space for personal choice, freedom, and agency. While they disagreed across the board which specific beliefs, values, and practices belonged in each of the domains, whichever domain they placed them into strongly predicted how they responded to divergence.[6] Specifically, if they located something in the personal domain, then they viewed the divergence as completely acceptable – people should be free to do as they wish (within reason), to make choices for themselves, and exercise their unique individuality and creativity.[7] If something was social/conventional, on the other hand, then they viewed divergence as less acceptable – people must be responsive to, and respectful of, the norms set by the group (e.g., you don't talk in class without raising your hand if the norm requires you to raise your hand first). That said, if someone violated a social/conventional norm, this was viewed as inappropriate, but not a sign of deviance. However, if the divergence was located in the moral domain, then such violations became highly unacceptable. They were viewed as forms of deviance, and those who committed them were to be avoided, if not shunned.

This is in line with decades of other research, which found that people treat moral transgressions as more serious, less permissible, less response-dependent, more severely punishable, and more universally generalisable than social/conventional wrongs (Turiel, 1983, 1998; see also Davidson, Turiel, & Black, 1983; Goodwin & Darley, 2008; Nichols, 2004; Nichols & Folds-Bennett, 2003; Nucci, 1981; Smetana, 1981, 1983; Smetana & Braeges, 1990), and they view moral transgressions as wrong even in the absence of rules and/or in the presence of social sanction (Smetana, 1981, 1983; Stoddart & Turiel, 1985; Turiel, 1983).

Meta-ethical pluralism

The fact that viewing divergent beliefs, values, and practices as moral so effectively shuts down tolerance for them – thereby protecting the socio-cultural group from deviance – introduces a problem. Once something is labeled as *moral*, this eliminates the space for conversation, debate, and disagreement; it's no longer viewed as diversity, but instead as deviance (Haidt, Rosenberg, & Hom, 2003), and so there is no longer room for individual choice.

But as morally imperfect beings, we cannot always be sure we have it right – and thus, we often run the risk of shutting down the conversation, of activating our intolerance, too soon. What if we label as *deviant* a belief, value, or practice that actually is not – and thereby prohibit someone from believing, valuing, or doing something that they should have the space to be able to believe, value, or do? The intolerance generated by the moral domain makes labeling something as moral *itself* potentially harmful (one of the arguments given by moral abolitionists for why we should stop thinking in moral terms altogether; see Wright & Pölzler, 2019).

Yet, the consequence of *not* labeling deviance as moral – of not shutting it down before it takes hold – is also risky. Perhaps by failing in our vigilance, we will allow something genuinely harmful, unjust, or oppressive to become a part of our shared system of socio-cultural norms. And, as our long history of unjust and harmful social practices can attest, once something is a part of our social-cultural space, once it viewed as an enforceable norm, it can be very hard to change – despite the obvious harm, injustice, or oppression it manifests. One psychosocial strategy that we have argued people use to navigate this treacherous territory shows up, not in the labeling of something as moral, per se, but in meta-ethical grounding they attribute to it. More specifically, people vary in whether they view moral issues as being objectively, or non-objectively, grounded – which is, in turn, predictive of how they respond to divergent moral beliefs, values, and practices (Pölzler & Wright, 2019, 2020; Wright, 2015, 2018a; Wright, Grandjean, & McWhite, 2013; Wright, McWhite, & Grandjean, 2014).

Philosophers have long debated whether people are moral objectivists – that is, whether people view moral facts as objectively determined (i.e., not simply determined by an individual's or group's beliefs, values, customs, etc.) and whether our judgments about them are "truth-apt" (see Blackburn, 1984; Brink, 1989; Mackie, 1977; Smith, 1994). If people are objectivists, then they view moral beliefs, values, and practices as (objectively) right or wrong. And, thus, to the extent that someone else has divergent beliefs, values, and practices, she is *objectively* wrong, misguided, even dangerous and immoral (e.g., Rachels & Rachels, 2009; Wong, 1984, 2006; Wright & Sarkissian, 2012; cf. Snare, 1992).

In a series of studies conducted by my lab, we have shown that people are not *straightforwardly* objectivists – but they are not straightforwardly non-objectivists (e.g., subjectivists, relativists, etc.) either. Instead, people are typically a mixture of both. That is, they ground *some* of the beliefs, values, and practices they have labeled as "moral" objectively – viewing them as determined by objective, universal, and fundamental facts that exist independently of people's beliefs, preferences, attitudes, norms, or conventions. But then they ground *others* as non-objective; as determined by an individual's, or her socio-cultural community's, beliefs, preferences, attitudes, norms, or conventions (Pölzler & Wright, 2019; Wright & Sarkissian, 2012; Wright, 2015, 2018a; Wright et al., 2013, 2014).

When we first discovered this pluralism (which has since been replicated many times), we were not sure what to make of it. It could have been that people were simply confused – meta-ethics is not something people think about or are familiar with, after all. But the qualitative data we have collected, asking people to explain their classifications and judgments, alongside the findings themselves, strongly suggests that this is not the case (Wright et al., 2013). Indeed, people seem to understand the distinctions (at least in layman's terms) surprisingly well.

The evidence suggested instead that meta-ethical pluralism was performing a function; it provided a way to introduce an issue into the moral domain without the hazards of shutting down the conversation entirely. In other words, it allowed people to entertain the idea that a particular belief, value, or practice was *moral* without having to (at least initially) adopt the many mechanisms of intolerance that typically followed. In support of this view, we found that the divergent moral beliefs, values, and practices that people grounded non-objectively were treated with significantly less intolerance (measured in the same ways previously discussed) than those they had grounded objectively. People were significantly more willing to privately and publically support – and less inclined to shun or condemn – moral divergence when they viewed that divergence as being non-objectively grounded. They also viewed others with divergent moral views less hostilely, and were more willing to interact with and help them, when they grounded that moral divergence non-objectively than when they grounded it objectively (Pölzler & Wright, 2019, 2020; Sarkissian, Parks, Tien, Wright, & Knobe, 2011; Wright, 2015, 2018a; Wright et al., 2013, 2014).

Interestingly, we also found that whether people provided objective or non-objective groundings was strongly predicted by the level of consensus about the issues under consideration that they expected within their socio-cultural group. The more divided they viewed their socio-cultural group to be about a particular belief, value, or practice – such as abortion or euthanasia – the less likely they were to ground it objectively, even though they viewed it as moral. This suggests that providing divergent moral beliefs, values, and practices with a non-objective grounding recognises and maintains critical "normative space" for people to disagree about an issue, at least until a sufficient level of consensus is reached – at which point, the belief, value, or practice in question (such as slavery, or sex trafficking of children) can be safely given an objective grounding and no longer tolerated.[8]

In other words, the view being argued for here is that meta-ethical pluralism serves an important "psychosocial" function by modulating the level of permissible choice and dialogue about divergent moral beliefs, values, and practices, providing the normative space required for open discussion and disagreement. It allows people to acknowledge the moral significance of the belief, value, or practice under consideration – that is, to acknowledge that it is not simply a personal or social/conventional matter – without yet requiring (or even being allowed to require) the censorship, prohibition, and intolerance that typically accompanies the moral categorisation. It holds open the normative space for people to consider what is at stake, to engage in social debate and dialogue – to achieve a sort of collective reflective equilibrium – before granting its moral status full (objective) moral weight.[9]

Of course, even if this is correct, unless a reasonable philosophical defence of pluralism can be given, philosophers might object that an "error theory" of some sort may be warranted – namely, that people's moral discourse erroneously treats certain moral issues as non-objectively grounded. But, I've argued elsewhere (Wright, 2018a) that there is at least a *pragmatic* defense for people's meta-ethical pluralism, something akin to a form of "exculpatory pretense", such as was introduced to explain other linguistic practices (Turri, 2013). That is, perhaps people aren't pluralists "deep-down" but instead are using non-objectivism to create/maintain a space for choice, discussion, and debate – long enough, at least, for agreement between reasonable parties to be reached, at which point the issue's actual objective status is acknowledged or its moral status is abandoned.[10] Even so, such a practice is entirely consistent with (and justified by) the moral demands of civility (Calhoun, 2000), which requires, among other things, that collective spaces respect the moral autonomy of the individuals who occupy them – even when, as is common in a morally imperfect world, those individuals have morally divergent views.

As morally imperfect beings, it is likely that many of our current moral beliefs, values, and practices will turn out to be wrong, even damaging – and therefore need to be changed. Unfortunately, the socio-cultural and personal difficulty of rejecting previously held moral beliefs and adopting new ones is often great, since our moral beliefs occupy the very centre of what we take ourselves to value and who we take ourselves to be – without them, we would no longer be who we are (Strohminger & Nichols, 2014). Thus, changing our moral beliefs, values, and practices often requires major social/political/cultural (and personal) transformation.

Such transformations cannot happen overnight – we need sufficient "normative space" for people to successfully navigate disagreement and divergence. Thus, the capacity to view moral issues pluralistically may serve as a critical mechanism for these transformations to occur, allowing the boundaries of our moral concern to change. The gradual introduction of an issue into the moral domain (and into our collective moral conscience) is facilitated by our ability to separate the issue's moral significance from its normative grounding. As our research suggests, coming to view a particular issue as moral involves the recognition of its morally significant features (such as the fact that it causes harm, is unjust, or is oppressive), but coming to ground it objectively involves another step – namely, viewing those morally significant features as trumping all other considerations, shutting down divergent views.

Thus, the fact that domain classification and grounding can come apart creates an opportunity for people to acknowledge an issue as moral – by acknowledging the presence of morally relevant features – without facing the immediate risk of social (and personal) censorship and punishment, allowing attitudes to change gradually. Once the moral status of an issue has become more universally recognized (i.e., a certain level of consensus has been achieved), then a shift from relative to objective grounding may occur, at which point, higher levels of censorship and punishment will be introduced.[11]

Importance of virtue

Thus far, we have discussed two separate, but related, psychosocial processes that people appear to use to navigate normative space, balancing between the need to protect the socio-cultural group from deviance, yet at the same time maintain open space for diversity – and for the possibility of changing and improving existing beliefs, values, and practices; correcting for error and corruption.

But what role does virtue play in all this? Elsewhere (Wright, 2008, 2018b), I have argued that not only can we gain moral knowledge by becoming part of a socio-cultural group – by learning which beliefs and values to have and which practices to engage in – but this very process of absorbing and imitating existing beliefs, values, and practices typically aids in the development of the very capacities, dispositions, and traits (such as virtues) we need to critically analyse, reflect upon, improve, and even replace those beliefs, values, and practices.

In other words, becoming an active participant in a shared system of socio-cultural norms (adopting the beliefs, values, and practices of that group) creates meaningful opportunities for further moral growth and advancement – i.e., the discovery and development of new moral knowledge and/or the extension of existing moral knowledge to new beliefs, values, and practices. And it does so, in large part, through virtue.

To see this, consider: as imperfect moral beings, learning what the right thing to do is, and how to do it, is a gradual – perhaps even a life-long – process. Indeed, we arguably have to learn most of what we need to know about morality, including how to actually *be moral*. And one of the central ways we do this is through participating in the daily activities of our socio-cultural group, by watching and imitating other members as they communicate the relevant beliefs and values and engage in the relevant practices. Watching, listening to, and imitating others helps us to learn – for example – how others are to be treated and how responsibilities, resources, and burdens are appropriately shared by members of the socio-cultural group of which we are a part.

At the core of learning to be moral are the virtues. In other words, what members of our socio-cultural group are teaching us when they communicate various beliefs and values and engage in various practices is what it looks like to be moral – to be *respectful*, *brave*, *loyal*, *honest*, *compassionate*, *generous*, *patient*, and so on. We regularly check in on our elderly neighbours because to do so is compassionate, we stand up for a friend who is being unfairly treated because to do so is brave and loyal, we share what we have with those less fortunate because to do so is generous, and so on.

What we learn by observing and participating in the daily life of our socio-cultural group is what virtue – respect, bravery, loyalty, honesty, compassion, generosity, patience, and so on – *looks like*, as well as when and how they are to be displayed, by and to whom. And because what we observe is not isolated, not the idiosyncratic beliefs, values, and behaviors of a few select individuals, but rather repeatedly heard and witnessed across a wide range of people and situations, we come to recognise them as a shared system of beliefs, values, and practices that is held by the entire socio-cultural group – things everyone agrees are good to believe, value, and do.

Say, for example, that everyone in a socio-cultural group engages in a common greeting practice (e.g., handshaking) because they understand that to do so is to *be respectful* to others – something that is valued enough by the group to have become a common practice. While some members of the group may engage in these practices – may desire to be respectful – simply because that is what is expected of them (and they don't want to disappoint or risk offending others in their group), often there is something deeper going on.

When we learn to be respectful by greeting people in a certain way, we learn more than just that there are certain ways of being respectful. We also learn that it is *because* they are ways of being respectful that they are commonly practiced – that ways of being respectful are *valued* by the socio-cultural group. And this implies that we, too, as members of the group, should not only engage in them (because they are valued, and because we are thus expected to), but also that we should likewise value them.

And if we do certain things *because* they are respectful – and we have come to value being respectful – this provides internal motivation to learn/discover new ways of being respectful. Thus, while it may be through a *particular* set of expressed beliefs, values, and practices that we learned how to be respectful, this knowledge, once imparted, is nonetheless not restricted to the set of expressed beliefs, values, and practices from which it was gained. Instead, we recognise that there are a range of different ways of being respectful – all of which matter only insofar as they are *ways of transmitting respect.*

To sum up, there are two important things to note. First, that in learning the particular ways of being respectful shared by their socio-cultural group, people are learning something deeper – namely, that respect (and being respectful) is something valued by their socio-cultural group, something that they (as members of that group) should also, therefore, value. And second, once they have come to value being respectful, it is *this* (respect) that ultimately matters to them, not the specific ways in which it is transmitted. And since these shared beliefs, values, and practices matter only insofar as they are ways of being respectful, this opens people up to two important possibilities:

- First, to the possibility that there are many different ways of being respectful, some of which may as yet be unknown, and
- Second, to the possibility that existing patterns of being respectful may actually *not* be respectful (at least, not fully) – thereby opening the door to improvements, revisions, and replacements (for an example, see the discussion of the practice of men holding doors open for women in Calhoun, 2000).

Arguably, this serves as the foundation for the development of virtue, because in learning particular ways of being respectful, people also learn *what it is like* to be respectful. And to the extent that being respectful creates positive feedback loops – e.g., in treating people with respect, we feel their appreciation and gratitude, and it also feels good being respected in return – we come to not only understand, but also to *experience* the value of being respectful, thus becoming motivated to continue doing so, not just because others expect us to, but because we value *respect*, and *being respectful, itself.*

Thus, insofar as these positive feedback loops link up our beliefs, values, and practices to our internal motivations, they provide the necessary (though not sufficient) "kick-starter" for the development of virtue, which provides not only the capacity, but also the motivation, required to critically analyse, reflect upon, improve, and even replace our existing beliefs, values, and practices in order to ensure that they are indeed respectful.[12]

Take the example given by Annas (2011) of a boy who learned to be brave by witnessing a parent chasing off a dog. In her example, the boy witnessed his parent's behaviour and learned that this particular action (the chasing off of a dog that posed a threat to the child) was brave. But this is not all he learned – he also learned that his parent's action as a *good* thing, in the sense that it kept him safe and protected. And he may have recognised (either then or later) a number of similarities between that particular action and all the other ways his parents, and others, keep him safe and protected, which eventually generalises into all the ways that some members of his socio-cultural group keep other members of the socio-cultural group safe and protected.

In other words, he would eventually come to realize that *ways of being brave* are fairly widespread in his socio-cultural group and occur with some regularity, whenever situations in which it is called for arise – that *bravery*, as expressed through a shared pattern of beliefs, values, and practices, is valued by his socio-cultural group. As a member of his socio-cultural group, he would thus likely come to also want to *be brave* whenever it was called for. And he may even eventually, through various positive feedback loops – e.g., the feeling of pride when standing up for a friend who's being bullied – come to value bravery not simply because it is valued by his socio-cultural group, but because it is valuable *in its own right*, insofar as it helps to keep members of his socio-cultural group safe and protected from harms, injustices, and oppressions.

Further, the more he recognises (and experiences) the value of being brave, the more motivated he will become to evaluate those beliefs, values, and practices intended to express, promote, and display bravery (to determine whether they do so, and if so, how), as well as to explore new forms of bravery, ways of being brave that he has not yet witnessed, that are not currently a part of his shared patterns of beliefs, values, and practices. All of which could ultimately lead to the revision of those beliefs, values, and practices and/or the introduction of new ones, thereby expanding his understanding of what it is to be brave, fed by his desire to be so.

More generally, as people become oriented towards virtue – being respectful, brave, compassionate, honest, generous, patient, and so on – this provides an essential foundation for our shared systems of norms, serving as a critical safeguard of our socio-cultural space, as a mechanism through which the balance between diversity and deviance can be productively maintained. Among other things, it becomes clear that any belief, value, or practice that expresses, promotes, or displays virtue (especially those virtues most valued by the socio-cultural group) should be allowed to flourish – even when it diverges from existing beliefs, values, and practices. Likewise, any belief, value, or practice (including existing ones) that threatens, impedes, or acts against virtue should be discouraged, if not prevented altogether.

In other words, to return for moment to the social domain theory, virtues collectively help us to successfully navigate and regulate the relationship between the three domains. They do this, first, by helping us gain moral knowledge, allowing us to better understand the critical difference between genuine deviance and diversity; and second, by helping us better regulate our motivation to effectively act on that knowledge – to appropriately protect ourselves and others from pernicious injustices and harms, while at the same time safeguarding the space for individual agency and creative expression.

Of course, as morally imperfect beings, perfect balance is not achievable and there will inevitably be moments when the socio-cultural system is leaning too far in the direction of promoting diversity, at risk of introducing deviance, or too far in the direction of protecting against deviance, at risk of choking out diversity. In each of these cases, different virtues may become essential for correcting the imbalance. In the former case, those virtues oriented towards protective/conserving become essential – i.e., those that promote an appreciation for existing beliefs, values, and practices, that caution against too much change too quickly (e.g., loyalty, patience, temperance, etc.), while in the latter case, more corrective/transforming virtues may be more strongly called upon – i.e., those that push against existing beliefs, values, and practices, and call for change (e.g., bravery, honesty, etc.). Thus, which virtues are most essential for promoting and protecting the balance between opening the gates to diversity and holding the line against deviance will vary, not only across socio-cultural groups, but also across different time periods, as well. But for all socio-cultural groups, the cultivation of virtue through shared beliefs, values, and practices is essential for the healthy functioning of the group and the individuals living within it.

Notes

1 Of course, social domain theory is not meant to be a theory of what normative authority *actually* is or where it is located, but rather a theory of how people view things, how they navigate this normative space.
2 For readers who may be wondering where virtue fits into all of this, the argument I will make later is that virtues help us to better regulate this process – that the development and expression of virtues increase our moral knowledge, thereby decreasing the risk of moral error, and protect against corruption. As such, it facilitates our ability to critically analyse and reflect upon our shared social/cultural space, allowing us to improve upon or replace any beliefs, values, and practices that are (or have become) unreasonably harmful. I will also argue that virtues develop and are expressed within the social/cultural space we have created, meaning that there will necessarily be localized expressions of otherwise broadly accepted virtues (courage, honesty, compassion, etc.) and localized prioritization of certain virtues over others.
3 I should also note that I do not view virtues as breaking down into these same three domain categories (e.g., moral, social, and personal virtues) – instead, I think different virtues (which are, broadly speaking, either moral or epistemic) are most prominent in different domains. So, for example, civility may be most important in our social/cultural space, while grit may show up mostly in the personal domain (thank you to Kristján Kristjánsson for this question).
4 That is, when people were asked to explain why they classified beliefs, values, and practices into one of these domains – and what they thought these domains meant – they provided explanations that matched very well with the social domain theory model (Smetana, 1981, 1983; Smetana & Braeges, 1990; Turiel, 1983, 1998).
5 As an example, people typically locate the belief that ice cream is delicious, and the practice of playing sports, in the personal domain, while they locate the belief that people should wear business attire to a corporate board meeting and the practice of driving on the left side of the road in the social domain, and they locate the belief that women are inferior and the practice of trafficking children in the moral domain.
6 While the percentage of the variance in people's responses predicted by their domain classification varied, it was often as high as 75–85%. This means that 75–85% of the differences in their tolerance for divergent beliefs, values, and

practices was *explained by* the domain within which they believed that divergence to be located.

7 It is important to note the "should" here – i.e., that our capacity for choice and agency is considered a *protected* space. Elsewhere (Wright, in press), I argue that the personal domain is not actually separate from the moral domain, but is instead part of it, involving those beliefs, values, and practices that are *morally* protected from undue social/cultural control or intrusion, providing space for people's agency to thrive.

8 Our research suggests that objectivism and consensus go hand in hand – the more perceived consensus, the more likely people are to give an objective grounding, and objective groundings are more likely to receive high consensus ratings. Additionally, one of our studies (Wright et al., 2014) suggested one way in which these two things work together to produce intolerance – namely, that consensus mediates the relationship between objectivism and intolerance. In other words, when people provided an objective grounding for a moral issue, they believed that a high percentage of their socio-cultural group would agree with them, and their consensus ratings predicted their level of intolerance. This suggests one manner in which a belief, value, or practice shifts from diversity to deviance is it becomes some an "outlier". When you believe everyone else in your group believes and values the same things, or does the same things, as you, then the person who believes, values, or does otherwise is an oddity, an outlier – and should, therefore, be considered deviant.

9 Importantly, it also has the potential to do the opposite – to help people to move a particular belief, value, or practice (such as same-sex marriage) *out of* the moral domain. Giving it a non-objective grounding allows people to hold onto their moral convictions about a particular belief, value, or practice, while at the same time creating room for discussion and alternative beliefs, values, and practices – with the hope that at some point the issue will shift out of the moral domain altogether.

10 If this view is correct, then the "relativism" being employed by pluralists would not be considered *genuine* relativism. But in a new project I have just started working on with Thomas Pölzler, we argue that there are good reasons to believe that meta-ethical pluralism – including *both* people's objective and relative judgments – is philosophically genuine and defensible.

11 For example, it may be that several issues currently sitting on the fringes of our collective moral conscience (e.g., vegetarianism, the use of non-renewable resources) are gradually making this transition from largely a matter of "personal choice" to a dominantly relatively grounded moral issue, and will then eventually to an objectively grounded moral issue – seen by future generations as being as embedded in the same moral bedrock that slavery and the sexual abuse of women and children now are, but once were not.

12 This linking up of our beliefs, values, and practices to our internal motivations through positive feedback loops arguably contributes to more than just the development of discrete virtues – it also contributes to the development of practical wisdom (*phronesis*) more generally, helping us to be able to critically evaluate both when the expression of a virtue is called for and also what expressing that virtue, in any particular instance, would look like (for more, see Wright, Warren, & Snow, in press; also Darnell, Gulliford, Kristjánsson, & Panos, 2019).

References

Annas, J. (2011). *Intelligent virtue*. New York: Oxford University Press.

Blackburn, S. (1984). *Spreading the word: Groundings in the philosophy of language*. New York: Oxford University Press.

Brink, D. O. (1989). *Moral realism and the foundations of ethics.* New York: Cambridge University Press.

Calhoun, C. (2000). The virtue of civility. *Philosophy and Public Affairs, 29*(3), 251–275.

Darnell, C., Gulliford, L., Kristjánsson, K., & Panos, P. (2019). *Phronesis* and the knowledge – action gap in moral psychology and moral education: A new synthesis? *Human Development, 62*(3), 101–129.

Davidson, P., Turiel, E., & Black, A. (1983). The effect of stimulus familiarity on the use of criteria and justifications in children's social reasoning. *British Journal of Developmental Psychology, 1*, 49–65.

Goodwin, G., & Darley, J. (2008). The psychology of meta-ethics: Exploring objectivism. *Cognition, 106*, 1339–1366.

Haidt, J., Rosenberg, E., & Hom, H. (2003). Differentiating diversities: Moral diversity is not like other kinds. *Journal of Applied Social Psychology, 33*, 1–36.

Killen, M., & Nucci, L. (1995). Morality, autonomy, and social conflict. In M. Killen & D. Hart (Eds.), *Morality in everyday life: Developmental perspectives* (pp. 52–86). New York: Cambridge University Press.

Mackie, J. L. (1977). *Ethics: Inventing right and wrong.* New York: Penguin Books.

Nichols, S. (2004). After objectivity: An empirical study of moral judgment. *Philosophical Psychology, 17*, 5–28.

Nichols, S., & Folds-Bennett, T. (2003). Are children moral objectivists? Children's judgments about moral and response-dependent properties. *Cognition, 90*, 23–32.

Nucci, L. (1981). Conceptions of personal issues: A domain distinct from moral or societal concepts. *Child Development, 52*, 114–121.

Nucci, L. (1996). Morality and the personal sphere of actions. In E. Reed, E. Turiel, & T. Brown (Eds.), *Values and knowledge* (pp. 41–60). Mahwah, NJ: Lawrence Erlbaum.

Nucci, L., & Turiel, E. (2000). The moral and the personal: Sources of social conflict. In L. Nucci & G. B. Saxe (Eds.), *Culture, thought, and development* (pp. 115–137). Mahwah, NJ: Lawrence Erlbaum Associates.

Pölzler, T., & Wright, J. C. (2019). Empirical research on folk moral objectivism. *Philosophy Compass, 14*, 1–15, https://doi.org/10.1111/phc3.12589

Pölzler, T., & Wright, J. C. (2020). Anti-realist pluralism: A new approach to folk metaethics. *Review of Philosophy and Psychology, 11*, 53–82.

Rachels, J., & Rachels, S. (2009). *The elements of moral philosophy.* New York: McGraw Hill.

Sarkissian, H., Parks, J., Tien, D., Wright, J. C., & Knobe, J. (2011). Folk moral relativism. *Mind & Language, 26*(4), 482–505.

Skitka, L. J., Bauman, C. W., & Sargis, E. G. (2005). Moral conviction: Another contributor to attitude strength or something more? *Journal of Personality and Social Psychology, 88*, 895–917.

Skitka, L. J., & Mullen, E. (2002). The dark side of moral conviction. *Analyses of Social Issues and Public Policy, 7*, 35–41.

Smetana, J. G. (1981). Preschool children's conceptions of moral and social rules. *Child Development, 52*, 1333–1336.

Smetana, J. G. (1983). Social-cognitive development: Domain distinctions and coordinations. *Developmental Review, 3*, 131–147.

Smetana, J. G., & Braeges, J. (1990). The development of toddlers' moral and conventional judgments. *Merrill-Palmer Quarterly, 36*, 329–346.

Smith, M. (1994). *The moral problem*. Oxford: Blackwell Publishing.

Snare, F. (1992). *The nature of moral thinking*. London: Routledge.

Stoddart, T., & Turiel, E. (1985). Children's concepts of cross-gender activities. *Child Development*, *56*, 1241–1253.

Strohminger, N., & Nichols, S. (2014). The essential moral self. *Cognition*, *131*, 159–171.

Turiel, E. (1983). *The development of social knowledge: Morality and conventions*. New York: Cambridge University Press.

Turiel, E. (1998). The development of morality. In N. Eisenberg (Ed.), *Handbook of child psychology: Social, emotional, and personality development* (Vol. 3, 5th ed., pp. 701–778). New York: Wiley.

Turri, J. (2013). The test of truth: An experimental investigation of the norm of assertion. *Cognition*, *129*, 279–291.

Wainryb, C., Shaw, L. A., Langley, M., Cottam, K., & Lewis, R. (2004). Children's thinking about diversity of belief in the early school years: Judgments of relativism, tolerance, and disagreeing persons. *Child Development*, *75*(3), 687–703.

Wainryb, C., Shaw, L. A., Laupa, M., & Smith, K. R. (2001). Children's, adolescents', and young adults' thinking about different types of disagreements. *Developmental Psychology*, *37*, 373–386.

Wainryb, C., Shaw, L. A., & Maianu, C. (1998). Tolerance and intolerance: Children's and adolescents' judgments of dissenting beliefs, speech, persons, and conduct. *Child Development*, *69*, 1541–1555.

Wong, D. B. (1984). *Moral relativity*. Berkeley, CA: University of California Press.

Wong, D. B. (2006). *Natural moralities: A defence of pluralistic relativism*. New York: Oxford University Press.

Wright, J. C. (2008). The role of moral perception in mature moral agency. In J. Winewski (Ed.), *Moral perception* (pp. 1–24). Cambridge: Cambridge Scholars Publishing.

Wright, J. C. (2012). Children's and adolescents' tolerance for divergent beliefs: Exploring the cognitive and affective dimensions of moral conviction in our youth. *British Journal of Developmental Psychology*, *30*(4), 493–510.

Wright, J. C. (2015). Meta-ethical pluralism: A cautionary tale about cohesive moral communities [commentary]. *Behavioral and Brain Sciences*, *44*, 1–158. https://doi.org/10.1017/S0140525X14001344,e163.

Wright, J. C. (2018a). The fact and function of meta-ethical pluralism: Examining the evidence. In T. Lombrozo, S. Nichols, & J. Knobe (Eds.), *Oxford studies in experimental philosophy* (Vol. 2). New York: Oxford University Press.

Wright, J. C. (2018b). Moral knowledge as know-how. In A. Zimmerman, K. Jones, & M. Timmons (Eds.), *Routledge handbook of moral epistemology*. New York: Routledge Press; Taylor & Francis Publications.

Wright, J. C. (in press). Morality as a regulator of divergence: Deviance or diversity? Special Edition of *Social Cognition*, titled *Morality as a hub: Connections within and beyond social cognition*.

Wright, J. C., Cullum, J., & Schwab, N. (2008). The cognitive and affective dimensions of moral conviction: Implications for attitudinal and behavioral measures of interpersonal tolerance. *Personality and Social Psychology Bulletin*, *34*(11), 1461–1476.

Wright, J. C., Grandjean, P., & McWhite, C. (2013). The meta-ethical grounding of our moral beliefs: Evidence for meta-ethical pluralism. *Philosophical Psychology*, *26*(3), 336–361.

Wright, J. C., McWhite, C., & Grandjean, P. (2014). The cognitive mechanisms of intolerance: Do our meta-ethical commitments matter? In T. Lombrozo, S. Nichols, & J. Knobe (Eds.), *Oxford studies in experimental philosophy* (Vol. 1). New York: Oxford University Press.

Wright, J. C., & Pölzler, T. (2019). *Should morality be abolished? An empirical challenge to the argument from intolerance*. Manuscript Under Review.

Wright, J. C., & Sarkissian, H. (2012). Meta-Ethics. In F. Allhof, R. Mallon, & S. Nichols (Eds.), *Philosophy: Traditional and experimental readings*. New York: Oxford University Press.

Wright, J. C., Warren, M, & Snow, N. (in press). *Understanding virtue: Theory and measurement*. Oxford: Oxford University Press.

Section 3

Practical and educational issues

For this section, we chose contributions that focused on the key question of the volume, about virtues as local or universal, by drawing on practical (especially professional) or educational examples and case studies. None of the chapters in Section 3 are "cheaply practical", however, by circumventing theoretical issues altogether. In contrast, the strength of the contributions here is that they weave together the warp of the theoretical with the weft of the practical – or, perhaps better put, they show the place of practical concerns within the greater fabric of theorising about universality versus particularity. Somewhat surprisingly (among academics who seldom agree on anything), there is significant concordance between most of the chapters in this section in that they mediate a middle-ground position and demonstrate the necessary nuances of responding satisfactorily to the previously mentioned key question without coming down firmly on either side of the "great divide". Aristotle's words about the aim of ethical inquiry being to do the good, rather than just to know the good, are often cited. He warns against those who "take refuge" in theoretical arguments, thinking that by so doing one is "doing philosophy" (1985, p. 1105b12–14). None of the authors here falls into that trap.

In Chapter 10, MacAllister discusses the complexities of Alasdair MacIntyre's position, which seems ambivalent in many ways regarding the universality of virtue and also constantly changing, as noted in our Introduction to Section 1. Despite these ambivalences, MacAllister senses some continuity in MacIntyre's grappling with our key question, and he takes this answer to be more persuasive in the end than Nussbaum's purely naturalist one, by valuing more the traditions, practices, and agency of local communities. He illustrates this conclusion with reflections on how moral education can empower students and help them think for themselves. MacAllister considers MacIntyre to provide a more realistic and balanced answer to the question of effective moral education than Nussbaum with his focus on moral education as a self-quest for the good life within community-bound narrative practices. Yet he acknowledges that if we want to make sense of the pursuit of modernist ideals of human individuality, freedom and authenticity as constituents of moral education, we may need to look beyond both Nussbaum's naturalism and MacIntyre's communitarianism.

In Chapter 11, de Ruyter contrasts the simple cultural homogeneity of her Protestant Dutch upbringing with life in contemporary, not only multicultural but "super-diverse", Western societies. After analysing the nature of the latter, she comes down on the side of the standard liberal answer that we need to respect the variety and diversity of different cultural practices while adhering to certain basic civic virtues and principles that cut across cultural divides. However, rather than just leaving the answer at the theoretical level, as most liberal thinkers do, de Ruyter elaborates upon a certain disposition that needs to be cultivated for the mediation of the universal and local to work. She calls it "the double gaze", and it has to do with both perceiving the unique otherness of the other *and* that person's common humanity. The double gaze "is thus both blind to difference as well as a precise vision of the richness of the identity of the other". She ends by offering various pedagogical suggestions about how teachers can both cover their subjects in a way that is conducive to the formation of the double gaze in students and also model it through their own attitudes and behaviours in the classroom.

In Chapter 14, Beadle and Robson explore, through a MacIntyrean lens, a practice that is in many ways unique and inimitable, but at the same time extremely illuminating regarding the universality-of-virtue question: that of the travelling circus. Their study reveals that the notion of locality of practice or virtue cannot be understood in mere geographical terms (for the circus is a moving feast), but is better understood in terms of particularity. The authors then use this case study to argue for what they consider to be MacIntyre's answer from *After Virtue* (1981): namely, that at least three virtues (justice, courage, and truthfulness) are truly universal, whereas others, such as temperance, are more particularised. For example, temperance is a matter of survival for flying trapeze artists, while relatively unimportant to clowns. Nevertheless, both need courage to survive in the profession, and the circus as a whole cannot thrive without the basic virtues of justice and truthfulness being adhered to within the practice. After presenting an array of illuminating qualitative data, Beadle and Robson conclude that the virtues of circus life are "to some degree universal and to some degree particular".

Two chapters deal with the question of virtues within the context of the profession of social work. In Chapter 12, Banks notes how this profession provides a particularly fruitful crucible in which to study local versus global virtues, as social work can be construed simultaneously as a locally based practice, aimed at empowering people to function better in their own unique environment, and a recognised international profession with a universal code of practice. Recent trends have aimed at deuniversalising and decolonising the language of social work and acknowledging that even if this language refers to virtues, those tend to be "thin" and open to different cultural interpretations. However, Banks is interested in the question of whether there also exist "thicker" profession-specific virtues that all social workers would be wise to adhere to, and she explores this question in the context of accounts written by the practitioners themselves, from over the world, of situations they found ethically challenging. She concludes that there is a common universal basis of recognition of some virtues, and that a mere surrender to allegedly incommensurable diversity is indicative of ethical failure: failure to understand others and engage with them through an open critical dialogue. Banks refers to the capacity to engage in such work as "professional ethical wisdom", a concept that leads nicely on to the other chapter about social work.

In Chapter 13, Weber and Wulfekühler explore the "translation work" that *phronesis* as practical wisdom can do in situations in which social workers experience conflicts, including conflicts between different cultural worlds. By building a bridge between a general blueprint of the good life and its irreducible particularity (Darnell, Gulliford, Kristjánsson, & Panos, 2019), phronesis makes the sort of dialogue that Banks referred to above possible and opens up "a learning process in which the alterity of the other is recognised". Through rich examples from actual practice, Weber and Wulfekühler illustrate the role that empathy, recognition and authenticity play in this learning process and how those relate to phronesis. Although this chapter, as well as the one by Banks, carries implications for social work education, Weber and Wulfekühler show how professional phronesis cannot be cultivated fully in any university course but must develop gradually through the self-reflection of the practitioner.

Finally, in Chapter 15, Chen offers a Confucian take on our key question. Although Confucius does not directly address the issue of culture-relative virtues, it is clear from his writing that the chief virtue of *ren* (benevolence, humanity) is not the privileged province of any superior category of people. Insofar as ideal character is concerned, for Confucius, then, there is no distinction between individuals based on their status, as in Aristotle. This appeal to universality is "part of the structural vision" of Confucian philosophy – making it even less amenable to a relativised interpretation than Aristotelian virtue theory. However, in Confucius, learning to follow rituals replaces the Aristotelian educational role of habituation in early years moral education. Other than that, there are significant philosophical and pedagogical similarities between these two ancient thinkers – providing further backbone to the impression to be gleaned from Section 3 of this volume that the great majority of thinkers in the virtue ethical tradition seem to share certain common assumption about what good character education involves and requires.

References

Aristotle (1985). *Nicomachean ethics* (T. Irwin, Trans.). Indianapolis: Hackett Publishing.

Darnell, C., Gulliford, L., Kristjánsson, K., & Panos, P. (2019). *Phronesis* and the knowledge – action gap in moral psychology and moral education: A new synthesis? *Human Development*, 62(3), 101–129.

MacIntyre, A. (1981). *After virtue*. Notre Dame: University of Notre Dame Press.

10 Educating virtues in a plural world

MacIntyre, Williams, and Nussbum on moral education

James MacAllister

Are virtues local or universal in nature?

Are the virtues needed for a flourishing life local or universal in nature? A philosophically satisfactory answer to this question must go well beyond simply stating that the virtues in question are "local" or "universal". I certainly do not though believe there is good reason to think that the specific character traits that mark *flourishing lives* are, always have been, and always will be everywhere the same. It seems clear that different individuals and cultures have taken the excellent life to mean different, and sometimes radically different, things. Alasdair MacIntyre (1984) famously argued that conceptions of virtue have and will continue to alter over time. I think he is right. He also argued that in order to make sense of the nature of virtue, it is necessary to compile a history of the concept. In his history of the concept, at least two very different instantiations of virtue emerge. First, *Homeric virtue* which required flourishing individuals to take on the social role of excellent warrior or athlete. Here, the virtues needed for flourishing involve physical prowess, competitiveness, cunning, and fearlessness about death. Second are the virtues of *Aristotle*, *Aquinas*, and *The New Testament*. In this adaptation of the concept, virtue is a quality that enables individuals to move toward their specifically human telos in this life or the next. The virtues needed for flourishing here (at least for Aristotle and Aquinas) centrally involve the rational ordering of activities, goods and relationships that are worthwhile in themselves. If the different virtues of Homer, Aristotle and the New Testament are considered, MacIntyre states "it would be all too easy to conclude that there are a number of rival and alternative conceptions of the virtues, but . . . no single core conception" (1984, p. 181). MacIntyre does not, however, draw the all-too-easy conclusion. Instead, he argues that a core unity to the concept of virtue can be found and that understanding the history of the concept is necessary to finding this unity (1984, p. 186). Though there is a unity to be found in the concept of virtue, the unity in question is not for MacIntyre based upon universal aspects of the human condition that are constant irrespective of time and place.

MacIntyre states "there are in no way universal features of the human condition" (1984, p. 67). Instead, he suggests the concept needs to undergo three phases of development before the unities in virtue can be understood. First, it needs to be recognised that *practices* are the primary (but not the only) arena in which virtues can be developed and exercised. Practices refer to well established, social human activities that have standards of excellence internal to them. The agent who pursues a practice in such a way that they learn to value that practice in its own right and not as a means to something else like money or status or power is, in effect, pursuing that practice virtuously. They pursue the practice with focus on the standards of excellence internal to that practice rather than for the sake of goods external to the practice. In doing so, they are going some way to leading a flourishing life. However, flourishing second requires comprehension of the fundamentally narrative and teleological character of human life. Human beings are, as MacIntyre famously put it, "story telling animals" not essentially but through their history. While life is always partly unpredictable, and while people are at most only co-authors of their own lives, there is nonetheless for MacIntyre a narrative unity to be found in each human life. This narrative unity takes the form of a telos or quest for the good. This quest for the good "is always an education" (MacIntyre, 1984, p. 219). Virtues are needed to not only sustain the health of practices, then, but also to help people in their quest for their good. Third, MacIntyre suggests that what the existence of different ideas about virtue like the Homeric and Aristotelian show is not that virtue is always at base culturally relative, but that different embodiments of virtue belong to a moral tradition of virtue – a tradition which, if in good order, will embody continuity and conflict over time. MacIntyre's suggestion that concepts of virtue differ, and evolve over time, does have a certain relativistic hue. However, MacIntyre resists the idea that the virtues are irrevocably local or relative in nature. Instead, he suggests that if the different versions of virtue that have come in to being are viewed with historical awareness, certain unities in the concept emerge. These unities are threefold: (1) *practices* are the primary place in which virtues are developed and exercised; (2) each human life has a narrative unity brought together by a teleological quest for the good; and (3) the tradition of virtue needs to embody "continuities of conflict" if that tradition is to continue to sustain people in their quest for the good. This interpretation differs from a standard relativistic reading of virtue in that according to the latter, moral standards, habits, and values are at base always relative to a person or group. For the relativist, there are no universal moral facts or standards of conduct that apply to all. What is virtuous in one context need not be in another.

In his most recent work, *Ethics and the Conflicts of Modernity*, MacIntyre pushes back further against ethical and cultural relativism. There he maintains there can be a "large measure of agreement" (2016, p. 223) about the sorts of virtues needed for flourishing. He also claims it is hard to deny that at least eight different goods can be identified that can contribute to a good life irrespective of one's culture. These are:

(1) Good health and a standard of living that is free of destitution
(2) Good family relationships
(3) Sufficient education to develop one's powers
(4) Productive and rewarding work
(5) Good friends
(6) Leisure time for athletic, aesthetic, and intellectual activities good in themselves
(7) The ability to rationally order one's life
(8) The ability to learn from one's mistakes

(MacIntyre, 2016, p. 222)

MacIntyre insists that it is possible to lead an excellent life without one or more of these goods. However, he stresses that the more of these goods that are absent, the more resourceful an agent will have to be to compensate for the absence. Importantly, the resourcefulness needed involves a capacity to see what requires alteration in either the agent or the agents wider social and institutional order so that the agent can achieve goods characteristic of an excellent life. MacIntyre suggests that Items 3 and 7 are especially important if agents are going to be able to deliberate well and make choices that enable their flourishing. He states that it "is by their initial education as practical reasoners and by their subsequent exercise of their reasoning powers in the making of such choices that agents play their part in determining the goodness of their lives" (MacIntyre, 2016, p. 223). MacIntyre does not renounce his earlier denial of there being universal features of the human condition or his suggestion that there is a unity to be found in different concepts of virtue. He only says it is hard to deny that these eight goods will likely support human flourishing, whatever social and political order one belongs to. He does not say the eight goods are universal features of the human condition, per se. MacInyre formulated the eight goods above in response to criticisms levelled against Aristotelian ethics by Bernard Williams. Indeed, MacIntyre remarks that some of the most pertinent objections against Aristotelian ethics (and thus his own) come from Bernard Williams. It is to Williams that I now turn.

Two problems for Aristotelianism in *Ethics and the Limits of Philosophy*

> Aristotle saw a certain kind of ethical, cultural and indeed political life as a harmonious culmination of human potentialities, recoverable from an absolute understanding of human nature. We have no reason to believe in that.
>
> (Williams, 2011, p. 59)

In *Ethics and the Limits of Philosophy* (2011, originally 1985), Bernard Williams maintained that in the modern world, there is no good reason to believe the Aristotelian story about how the dispositions of the virtuous agent are founded in universal facts about human nature. Williams also claimed that Aristotle should not have believed that human beings are responsible for their character. It is Aristotle's account of habituation which Williams finds particularly problematic. It provides the foundation for a person's character yet "leaves little room for practical reason to alter radically the objectives that a grown-up person has acquired" (Williams, 2011, p. 44). According to Williams, as Aristotle thought the virtues can only be acquired through habituation in the early years, then the practical reasoning of adults will quite simply come too late for any fundamental reformation of character to be possible. Habituated persons are already "preemptively good or irrecoverably bad" (Williams, 2011, p. 44). It is the *irrecoverably bad* person that Williams is especially concerned with. Williams insists that Aristotle did not say enough about why it is in the real interests of the person of bad upbringing to be virtuous (Williams, 2011, p. 50). What he is getting at is that the unrepentant person of bad upbringing presents a challenge to a core aspect of Aristotle's ethics – that human beings need to lead a life of virtue in order to flourish. The challenge concerns how to convince a person of bad upbringing, say the career gangster brought up in to a life of crime from a young age, to realign their desires, goals, and doings towards what is really and not mistakenly in their interests. Williams thinks this is a challenge Aristotle's ethics is not well placed to solve.

On Aristotle's account, what is really in a person's interest is that which conduces to their long-term flourishing. MacIntyre explains that an:

> Aristotelian theory of virtue does therefore presuppose a crucial distinction between what any particular individual at any particular time takes to be good for him and what is really good for him. . . . It is for the sake of this latter good that we practice the virtues.
>
> (1984, p. 150)

Williams and MacIntyre agree that Aristotle thought people may choose to pursue objects that they perceive to be in their interests when in point of fact these objects will not really be good for them in the specific sense of *good for their long-term flourishing*. Williams held that agents could not be motivated to act by reasons that are external to their subjective motivations. According to Williams (1981), an agent has an *internal* reason to perform an action if that action will help a subjective desire of theirs to be satisfied. Agents lacking subjective motivation to do something have no internal reason to do it. Williams appears, then, to be suggesting this: that adults who have been habituated to pursue what Aristotelians would take to be mistaken interests, have no *internal* reasons to act in ways that will aid their flourishing, and that this may be through no fault of their own. They were not responsible for their habituation so they should not be held responsible for the resultant patterns of desire, thought and action established by habituation.

Williams acknowledges how Aristotle's ethics can help strengthen the resolve and deepen the insight into the ethical life of those already "inside" it. Williams also credits Aristotle with a more true-to-life portrait of the "bad" character than Plato. While Plato tended to portray the vicious person as a compulsive addict lacking in rationality, Aristotle saw that the vicious person could use their reason to help them attain whatever vicious ends they sought. However, this more realistic portrait of the vicious person presents a problem for Aristotle's normative theory – why should the person habituated into vice pursue a life of virtue? Williams had another related reason for rejecting the justification of morality provided by Aristotle's ethics – its foundation in a teleological view of human nature. Williams (2011) argued that Aristotle was not very successful in demonstrating how the telos of human nature could be harmoniously brought to fruition through the life of virtue. This argument in no small part hinges upon how Williams thought contemporary developments in psychology and evolutionary biology, do not support Aristotle's conclusion that an ethics of virtue and flourishing is part of human nature and human purpose. In sum, Williams presents at least two challenges to Aristotelian ethics. First, Aristotle did not provide a satisfactory theory of error about how people of "poor upbringing" can fail to recognize their real interests. Second, in light of the findings of modern science, it is no longer plausible to accept the connection between the facts of nature and ethical considerations that Aristotle's teleological view of human beings requires. I will now draw upon first MacIntyre and then Nussbaum to show how Williams' criticisms of Aristotle have been responded to.

On the importance of learning from ethical error

MacIntyre (2016) addresses Williams' charge that Aristotle fails to provide a convincing theory of ethical error. He suggests there are two possible ways in which to rebut Williams. One could first document the various moral and political errors Aristotle does delimit – the mistakes of the akratic or intemperate person, or of those communities like the Spartans who excelled at war but were intellectually limited. However, I do not think Williams would likely be convinced by this line of argument. All these examples illustrate is that Aristotle did describe some ways in which communities and persons could ethically err. They do not explain why it is in the real interests of the person or community that ethically errs to change their ways. However, MacIntyre's second line of rebuttal more directly confronts the issue of real interests and ethical error and is more promising for it. According to MacIntyre, ethical error involves an agent desiring what they do not have good reason to desire. MacIntyre argues that in Aristotelian ethics, it should not be assumed that the source of moral and/or political error must be the *individual* moral agent. Instead, the source of error could be either in the individual agent or in the social relationships that influence or have influenced the individual moral agent. Why is this distinction important?

It is important because it calls in to question Williams' assertion that Aristotle should not have believed human beings to be responsible for their characters. What MacIntyre brings out is that on Aristotle's view, a teacher or parent may be at least as, if not more, responsible for the ethical errors of their student or child. However, the crucial issue for MacIntyre is not whom should be held accountable for ethical error, but what might be learned from ethical errors when they occur; in particular, what might be learned that will help agents become more able to desire what they have good reason to desire in the future. MacIntyre argues that the virtuous exercise of practical reason requires not only individual, but also shared, deliberation about what agents have good reason to desire. Contrary to Williams, he maintains that it is not just early habituation that makes a person desire what they do. Instead, human desires change over the course of life. Importantly, human desires can be redirected through reflection upon errors and through practices of shared deliberation. MacIntyre argues that moral agents can only learn how to reason well in their lives by engaging in "mutual criticism" with family members, friends, co-workers, or anyone else in their life with whom they share practical concerns (2016, p. 224). On Aristotle's account, agents do not reason about their desires and ends in isolation – they need to reason with others about their desires and ends. Such shared deliberation can help moral agents come to see that what they thought they had good reason to desire was not really in their long-term well-being interests after all. It is therefore my argument that as such shared practical reasoning can help agents identify errors in their desires and ends, in ways that can lead to agents desiring new ends that are in their real interests, Aristotelianism can provide a coherent theory of ethical error.

Nussbaum (1995) takes issue with Williams' second criticism, maintaining that Aristotle's ethical and scientific inquiries are not neatly comparable to what we would today recognize as science. When he in various places took a fact about human nature in order to defend an ethical conclusion, he was not adopting a modern scientific view of the human being. Instead, he took for granted that human beings are not beasts or gods or jellyfish. The human condition is one of both capability (the capability to be social and to lead a life of practical reason) and limitation (all humans have, for example, distinctively human bodies that are mortal). These capabilities and limitations are facts of nature that are *internal* to human beings and they are facts that shape what it means to lead a human life well. Nussbaum argued that Aristotle grounded his views on human nature and ethics upon beliefs, social traditions, and the background of myths and tragedies that all ancient Greeks were aware of. Myths that revealed how human beings, to be true to their nature (unlike gods and beasts), need to live social and political lives. Williams (1995) replied to Nussbaum. He appreciated the connection Nussbaum made between tragedies and Aristotle's views of human nature. He also conceded that he no longer wanted to claim that "the Aristotelian enterprise requires a 'top-down' derivation of ethical conclusions from a scientifically respectable account of human beings" (Williams, 1995, p. 200). Williams, however, thought Nussbaum's picture of human nature was in danger of not being representative of the range of different forms of human association that are possible. As we shall now see, Nussbaum has maintained that an Aristotelian account of human nature is compatible with plural forms of social life.

Nussbaum in defence of Aristotelian essentialism

> we do recognise others as human across many divisions of time and place. Whatever the differences we encounter, we are rarely in doubt as to when we are dealing with a human being and when we are not. The essentialist account attempts to describe . . . those features that constitute a life as human whatever it is.
>
> (Nussbaum, 1992, p. 215)

In *Human Functioning and Social Justice: In Defense of Aristotelian Essentialism*, Nussbaum maintains that an Aristotle-inspired, historically sensitive essentialism about human nature and human goods should inform international public policy. Such essentialism is warranted, she suggests, because it can help to better cater to the needs of many marginalised, vulnerable, and disenfranchised persons and groups in various different international contexts. In building this argument, she considers three common criticisms of essentialism. First, that essentialism is *neglectful of historical and cultural differences*. Here the charge is that human life is and has been actualised in widely differing ways at different times and places so that "any attempt to produce a list of essential human properties is bound to enshrine certain understandings of the human and to denote others" (Nussbaum, 1992, p. 208). Normally it is the understandings of the dominant group that come to the fore to the exclusion of the minority group. Furthermore, the coherence of any list of essential characteristics of all human persons will only be secure if there is unanimous agreement about the characteristics in question being possessed by all human persons. As no such agreement will be forthcoming in practice, essentialism is doomed to fail. The second objection to essentialism involves a *neglect of autonomy*. If the criteria of a good human life are determined in advance, any essentialist account of the human must by definition neglect the right of human persons to choose their own future and their own good in any ways that deviate from the pre-established criteria of a good human life.

The third objection concerns *prejudicial application*. If a determinate conception of the human being is defended, then it becomes all too easy to exclude the powerless and marginalized. Here, Nussbaum notes that

Aristotle's own essentialist account of human nature has been criticised on the grounds that women, slaves, children, and many productive workers were excluded from the political sphere as they were deemed to lack all the properties of the "fully" human. The "suggestion here is that renouncing the use of such a determinate conception of the human will make it easier for such people to be heard and included" (Nussbaum, 1992, p. 208). Nussbaum acknowledges each of these objections has some merit. However, she does not want to renounce essentialism. Why? Because she believes there is an essentialism that can "survive" these objections. What type of essentialism is this? How, moreover, does this essentialism survive these three objections? Nussbaum's overarching argument is that a historically sensitive and empirically informed essentialism is needed because it can enhance social justice. Nussbaum reasons that it is only after a set of especially important human functions have been identified that individuals and groups will be in a position to ask what social and political institutions are doing to support them to flourish. According to Nussbaum, the core functions of a good human life are:

(1) Being able to live to the end of human life
(2) Being able to have good health
(3) Being able to avoid unnecessary pain and have pleasurable experiences
(4) Being able to use the five senses and being able to imagine, think and reason
(5) Being able to have attachments to things and persons outside ourselves
(6) Being able to engage in critical reflection about the planning of one's life
(7) Being able to live for and with others
(8) Being able to live in relation with animals, plants and the world of nature
(9) Being able to laugh and play
(10) Being able to live one's own life

(adapted from Nussbaum, 1992, p. 222).

It is important to note that Nussbaum regards the features of the good human life as *capabilities*, as things that legislation and policy must seek to ensure for all human beings. Nussbaum also provides insight into how her Aristotelian theory of the human good can survive the three common objections against essentialism. In respect to the objection of *neglecting cultural differences*, Nussbaum believes that her list of common human properties is flexible enough to accommodate local and personal variations in lived practice. Indeed, she refers to this list as a "thick vague" theory of the good human life; vague as "it is better to be vaguely right than precisely wrong" (Nussbaum, 1992, p. 215). She insists that the Aristotelian essentialist about the good human life will want to retain plurality in two ways: via *plurality* and *local specification*. In terms of *plurality*, she indicates that her deliberately vague and general framework enables substantial scope for citizens to realise their own good "in accordance with local traditions or individual taste" (Nussbaum, 1992, p. 224). In terms of *local specification*, she rightly avers that Aristotelian practical reasoning is always carried out with deep sensitivity to the concrete local conditions in which the person or persons live. In regard to the objection of *neglecting autonomy*, Nussbaum points out that hers is a list of human *capabilities*, not actualities. As such, individual agents retain the right to choose whether or not to realise the human goods in question. The "*good*" human life does not therefore require that persons must develop the capabilities in question – only that they have the choice to. However, she points out that for Aristotelians, human choices are not highly individualistic or spontaneously arising. In much the same way that MacIntyre does, Nussbaum insists that human choices are shaped by the material and social conditions that persons live in. In regard to the final objection of *prejudicial application*, Nussbaum again demurs. She suggests that her list of capabilities is deliberately designed to be non-prejudicial insofar as it refers to all human beings – the crucial point being that in her framework, no person will be excluded from accessing goods on account of their being less than fully human.

MacIntyre on what makes human beings human

> Aristotle is . . . obviously someone whose arguments go badly astray. . . . His conception of the natural slave . . . and his claim that women are unlike men in their inability to control their passions as reason dictates are . . . wrongheaded . . . Modern Aristotelians – I think, for example, of Martha Nussbaum – have not found it difficult to excise these absurdities, while remaining Aristotelian in their account of the virtues.
>
> (MacIntyre, 2016, pp. 85–86)

MacIntyre and Nussbaum read Aristotle in some very similar ways. Neither philosopher takes Williams' arguments against Aristotelian ethics to be fatal. They both nonetheless think that Aristotle's ethics needs to be understood historically and be open to plural forms of social life if it is to continue to be relevant and defensible. They both relatedly think his idea that slaves and women lack the capacity of reason to be wrongheaded and in need of rejection. However, in spite of these similarities, the Aristotelianism of MacIntyre and Nussbaum is not identical. Nussbaum describes her Aristotelianism as essentialist (1992) and defends a "non-relative" (1993) theory of virtue. Informed by the belief that virtues are non-relative, Nussbaum argues in later work that policy makers have a responsibility to ensure that all people have opportunity to develop ten capabilities. Only if political states do this will they be minimally just (Nussbaum, 2011). MacIntyre does compile a list of eight goods he thinks it is hard to deny help human flourishing, irrespective of one's social order, but he does not insist flourishing is impossible without some or all of these goods. Nor does he argue that virtues are non-relative, and that policy makers in political states need to act to ensure conditions of minimal justice are satisfied. While he suggests that there can be a large measure of agreement across cultures about what virtues are needed for human flourishing, he also stresses that concepts and practices of virtue differ according to context, and have evolved and will continue to evolve over time. MacIntyre's Aristotelianism therefore strikes me as less essentialist than Nussbaum's. Indeed, MacIntyre's Aristotelianism has been described by Kelvin Knight (2013) as "revolutionary", not essentialist, and I think this is right. MacIntyre not only thinks objectionable features in Aristotle's outlook stand in need of revolution. He also thinks virtues are needed by agents if they are to learn how to challenge unjust social orders. The differences do not end here. MacIntyre is not convinced by the particulars of Nussbaum's response to Williams' claim that modern science renders Aristotelian teleology untenable. MacIntyre and Nussbaum do, to be sure, both think human beings have flourishing as the end of their actions, and that this is part of their nature qua human being – Williams does not. However, in order to rebut Williams, MacIntyre takes a different strategy to Nussbaum. MacIntyre seeks to explain how there is more to human nature and purpose than what evolutionary biology can tell us. In contrast to this, Nussbaum's main strategy was to insist that the "science" of Aristotle and the science of today are fundamentally incomparable, and that because of this we should not expect Aristotelian ethics to be relatable to the facts of modern science.

MacIntyre agrees with Williams, not Nussbaum, on this important point, though – that Aristotelian ethics requires commitment to some sort of thesis about how teleology can explain the facts of nature. Williams (1972) claimed that the identification of rationality as the characteristic aspect of human beings is arbitrary, and that falling in love is just as human a quality as rationality. In response to this, MacIntyre maintains that human beings can be differentiated from other species "by realizing possibilities that cannot be accounted for in solely evolutionary terms" (Williams, 1972, p. 226). These actualised possibilities include: having a concept of the good, being able to reason and order goods, spending longer in leisure pursuits that no longer confer reproductive advantage than other species, and by the amount and kind of education given to the young. MacIntyre does not dispute that falling in love is very human. However, he adds that educating our feelings and subjecting them to critical appraisal is also part of what makes us human. He says that "Williams was right in taking some human affective capacities to be distinctively human. Part of what makes them distinctive, however, is the ways in which they can be educated and their exercise criticised" (MacIntyre, 2016, p. 226). Unlike Nussbaum, MacIntyre shows how Aristotelianism can provide a theory of ethical error, suggesting that adults need to learn from their mistakes and continuously subject their actions, feelings, and choices to critical scrutiny. He stresses that we need others to help us work out what we have good reason to desire. Nussbaum and MacIntyre do nonetheless agree that it is vital to educate in such a way that ethical habits are formed, but whereby plural forms of human life and association are still valued. Interestingly, so does Williams.

Educating virtues in a plural world

In this chapter, I have considered two different responses to the question of whether the virtues needed for a flourishing life are local or universal in nature. First, MacIntyre, who suggests that if different instantiations of virtue are viewed with historical awareness, certain unities in the concept emerge. Second, Nussbaum, who maintains that virtues are non-relative, and that an Aristotle-inspired, historically sensitive essentialism about human nature should inform international public policy. Which response is best? I think MacIntyre's response is the most persuasive, for two reasons. First, in taking essentialist capabilities as the starting point for global policy and action, Nussbaum does not value the traditions, practices, and agency of local communities as much as MacIntyre does. Unlike Nussbaum, MacIntyre does not take a list of essential capabilities as a starting point for global debate, action, and policy concerning human well-being. Nor does he insist, as Nussbaum does (2011), that there is a list of ten human capabilities that political orders must ensure for all citizens if that region or country is to be considered minimally just. Instead, MacIntyre stresses ethical debate and action must begin from within each person's own particular social context. He states that people inherit from family, city, tribe, and nation "a variety of debts, inheritances, rightful expectations and obligations. These constitute the given of my life, my moral starting point" (1984, p. 220). MacIntyre does not, to be sure, endorse individualism or moral and cultural relativism. Instead, communities in good working order relate with and learn from other moral traditions so as to improve their own (1988). His account of the virtues has the advantage, then, of eschewing moral relativity while also being much less essentialist and transculturally meddlesome than Nussbaum's. Nussbaum (2011) does, of course, stress that her list of capabilities can be adapted by local cultures, but in stressing this, the implication is clear – capabilities ought to form the starting point and foundation of well-being policy and action, irrespective of place. Other cultures need to adapt to her list of capabilities. Nussbaum is thus more inclined to impose aspects of her Aristotelian ethics on others than is MacIntyre.

Second, I think MacIntyre more directly and convincingly addresses two key challenges posed to Aristotelian ethics by Williams. MacIntyre employed a different strategy than Nussbaum in response to Williams' second objection. He tried to explain how there is more to human nature and purpose than what evolutionary biology can tell us. While it is true that Nussbaum's key paper (1995) in response to Williams stressed that understandings and practices of science in the ancient Greek world are at base incomparable with those of today, she has elsewhere (1992) claimed that human beings need to achieve certain capabilities if they are to flourish. It might therefore be possible to argue that Nussbaum, like MacIntyre, provides an explanation about how the facts of human nature can be reconciled with teleological ethics. However, MacIntyre responded to Williams' first objection. Nussbaum has not done this, so far as I am aware. He sought to explain how Aristotelian ethics can provide a coherent theory of ethical error. MacIntyre acknowledges that Williams and those sympathetic to him would not be convinced by his theory of ethical error (MacIntyre, 2016, p. 221). This thought may not be entirely right, though. Williams, after all, maintained that "the formation of ethical dispositions is a natural process in human beings" (1984, p. 53). Related to this, he also maintained that as ethical life is sustained by the dispositions of people, "we have much reason for, and little reason against, bringing up children within the ethical world we inhabit" (MacIntyre, 2016, p. 54). In *The Primacy of Dispositions*, Williams goes further. He suggests that *only* a disposition view can generate a realistic account of ethical criticism, maintaining that if ethical life is to be changed, this will ultimately require changes in the dispositions of agents. Williams nonetheless doubted if Aristotelian ethics could value plural forms of human association or educate people in such a way that they could change the ethical dispositions of their habituation. He insisted, however, that institutions must value such association and education.

While Williams rejected a teleological account of human nature and purpose, and did not consequently agree with MacIntyre and Nussbaum at the level of *ethical theory*, there does appear to be at least some general agreement between them about the purpose of moral *education*. In particular, Williams, MacIntyre, and Nussbaum all suggest that ethical dispositions should be cultivated in education in ways whereby the young also learn how to subject the values of their culture to criticism. However, it would also seem important to acknowledge that Williams valued the pursuit of human individuality and authenticity much more than either MacIntyre or Nussbaum. Williams overtly questioned the idea that ethics should aim at some general or social well-being, and instead suggested that it might be more desirable to found morality on Lawrence's dictum to "find one's deepest impulse" and "follow that" (Williams, 1972, p. 79). While Williams wrote very little explicitly about education other than his paper on the primacy of dispositions (Williams, 1987), in view of the foregoing, it seems fair to conclude that he might well have wanted moral education to foster a greater spirit of individuality and freedom in the young than either Nussbaum or MacIntyre. In different ways, MacIntyre and Nussbaum do, however, show how any tenable Aristotelian ethics and education needs to value plural forms of human association. Nussbaum argues that education should cultivate humanity in ways where the plurality of human cultures is valued (1997). She defends a liberal university education, but argues that traditional academic disciplines on university curricula can and should be augmented by emerging disciplines. In her view, the study of newer bodies of knowledge, like women's studies and non-Western traditions of thought by students at Western universities, has potential to help such students learn to value plurality. Though MacIntyre and Dunne (2002) endorse having separate faith schools and universities, MacIntyre nonetheless stresses that it is crucial that students in such places of learning are not insulated from different perspectives and values. Instead, it is vital that they are exposed to different traditions of thinking and ways of living so that prejudices (which are for him an enemy of education) do not emerge.

MacIntyre and Nussbaum also stress the importance of educational institutions initiating students into their culture and social orders in such a way that students can learn how to think critically about them. Nussbaum (2002) draws a contrast between the *old* and *new* education that emerged in Athens at the time of Socrates, Plato, and Aristotle, indicating that educational institutions today should follow the latter "new" model. The old education entailed lots of memorising whereby young people were actively discouraged from questioning the teachings of their elders. In the *new* education, by contrast, young people learn to think critically about the social origins of their traditions and how to construct their own arguments. They will not just accept views of authority. They "know they need to use tradition to invigorate their own thought – but this benefit involves a willingness to criticize it when criticism is due" (Nussbaum, 2002, p. 294). Nussbaum suggests that a good liberal education will develop three capacities in students: the capacity to critically examine oneself and one's traditions, the capacity to think as a global citizen and not just as a citizen of a local region or group, and the capacity to imagine what it is to be in the position of someone very different (what she dubs the narrative imagination). For Nussbaum, reading literature from other cultures is one important way of fostering the narrative imagination – so too is fostering relationships with students (using for example Skype) from different parts of the world.

MacIntyre and Dunne (2002), in turn, suggested that education must aim to foster student powers and help students learn to think for themselves. For MacIntyre, thinking for oneself may often entail learning how to put difficult questions (about social injustices) to those in power (MacAllister, 2016). MacIntyre, like Nussbaum, also thinks that a balanced liberal education involving the study of different academic disciplines is needed if those of the next generation are going to learn how to respond with due thought and sensitivity to the most pressing social challenges of today. He maintains that teachers should help students extend their powers through participation in various literary, mathematical, scientific, musical, and athletic practices (MacIntyre, 2016, p. 173). The curricular recommendations that flow out from the ethical and educational theories of Nussbaum and MacIntyre have some marked similarities, then. There is also at least one important difference. Nussbaum's vision of education is thoroughly global and cosmopolitan – MacIntyre's is much more local. Nussbaum (1997) maintains that the fostering of global citizenship ought to be a key aim of education, whereas MacIntyre (2016) calls for parents, teachers, and local communities to have more say in how their schools are run. In taking capabilities as the starting point for international social and educational policy, Nussbaum does not value the traditions, practices, and priorities of local communities, as MacIntyre does. MacIntyre also insists, more so than Nussbaum, that adults can and should carry on subjecting their desires, practices, dispositions, and ends to critical scrutiny long after their formal education and childhood is over. Contrary to Williams, MacIntyre thinks Aristotelian dispositions can be changed when agents commit to learning from ethical errors.

Biography, narrative fiction, and film might be helpful in this respect. MacIntyre believes narratives to be of great moral and educational import. He remarks that human beings are tellers of stories "that aspire to truth" (1984, p. 216). We tell stories about ourselves in our quest for the good, and we need stories in order to properly understand the lives and actions of others. MacIntyre stresses that children without myths and fairy tales are deprived of a crucial resource that can aid thought, action, self-understanding, and understanding of their wider social orders (MacIntyre, 1984). While MacIntyre does not expressly maintain that those working in educational institutions ought to employ narratives of various kinds that show human beings learning from ethical errors in order to prompt student reflection about this, it is my contention that such pedagogical practices would be wholly consistent with his thinking on ethics and education. Human beings need narratives to help them consider how a life might be other than it is. Films like *Ikiru, Morven Callar*, or the more light-hearted *Stranger than Fiction* could, for example, be used in classrooms to foster discussion about how it is possible to learn from mistakes; change desires, habits, and course; and lead richer and more rewarding lives as a consequence. Well-chosen narratives can, in my view, be a powerful educational resource for prompting thought about the importance of learning from ethical errors in life. This is no small lesson.

References

Knight, K. (2013). Revolutionary Aristotelianism. In P. Blackledge & K. Knight (Eds.), *Virtue and politics: Alasdair MacIntyre's revolutionary Aristotelianism* (pp. 20–34). Notre Dame: University of Notre Dame Press.

MacAllister, J. (2016). MacIntyre's revolutionary Aristotelian philosophy and his idea of an educated public revisited. *Journal of Philosophy of Education*, *50*(4), 524–537.

MacIntyre, A. (1984). *After virtue: A study in moral theory*. Indiana: University of Notre Dame Press.

MacIntyre, A. (1988). *Whose justice? Which rationality?* London: Duckworth.

MacIntyre, A. (2016). *Ethics in the conflicts of modernity: An essay on desire, practical reason and narrative*. Cambridge: Cambridge University Press.
MacIntyre, A., & Dunne, J. (2002). Alasdair MacIntyre on education: In dialogue with Joseph Dunne. *Journal of Philosophy of Education*, *36*(1), 1–19.
Nussbaum, M. (1992). Human functioning and social justice: In defense of Aristotelian essentialism. *Political Theory*, *20*(2), 202–246.
Nussbaum, M. (1993). Non-Relative virtues: An Aristotelian approach. In M. Nussbaum & A. Sen (Eds.), *The quality of life* (pp. 242–269). Oxford: Clarendon Press.
Nussbaum, M. (1995). Aristotle on human nature and the foundation of ethics. In J. E. J. Altham & R. Harrison (Eds.), *World, mind and ethics: Essays on the ethical philosophy of Bernard Williams* (pp. 86–131). Cambridge: Cambridge University Press.
Nussbaum, M. (1997). *Cultivating humanity: A classical defense of reform in liberal education*. London: Harvard University Press.
Nussbaum, M. (2002). Education for citizenship in an era of global connection. *Studies in Philosophy and Education*, *21*, 289–303.
Nussbaum, M. (2011). *Creating capabilities: The human development approach*. London: Harvard University Press.
Williams, B. (1972). *Morality: An introduction to ethics*. Cambridge: Cambridge University Press.
Williams, B. (1981). Internal and external reasons. In *Moral luck: Philosophical papers* (pp. 101–113). Cambridge: Cambridge University Press.
Williams, B. (1987). The primacy of dispositions. In G. Haydon (Ed.), *Education and values: The Richard Peters lectures* (pp. 56–65). London: Institute of Education/University of London.
Williams, B. (1995). *World, mind and ethics: Essays on the ethical philosophy of Bernard Williams* (pp. 185–224). Cambridge: Cambridge University Press.
Williams, B. (2011). *Ethics and the limits of philosophy*. London: Routledge. (Original work published 1985).

11 Education and upbringing in a super-diverse society

On the cultivation of the double gaze

Doret de Ruyter

Introduction

The village in which I grew up in the 1960s and 1970s was ethnically homogeneous. Even though it was very close to Amsterdam, the population was predominantly white and of Dutch descent. In my primary school, for instance, there were only a handful of children from the former Dutch colonies. In the past 40 years, the village has grown twice its size and the high-rise flats are now inhabited by second- or third-generation guest labour families, asylum seekers from various African and Middle Eastern countries, and migrants from countries in the eastern part of Europe. My former primary school that is close to the flats is now populated by the children from those minority families, and only a handful of children from majority Dutch families. In the same decades, the indigenous Dutch population also became more diverse. For sure, although the village was ethnically homogeneous, it was culturally and religiously diverse. Dutch society at that time was what we called a pillarized society, being divided into four main groups, namely Protestants, Catholics, Liberals, and Socialists. These groups read the newspapers of their own pillar, watched the pillar's television programmes, children attended schools of their own pillar, and they all had their own political party. These pillars have fallen down; the Netherlands has become more individualistic, and individuals belong to various communities and have loose associations both in the real and the virtual worlds. Thus, Dutch society has become a culturally and ethnically diverse society.

Living together in a diverse society is not self-evident and needs to be learned. The question addressed in this chapter is which values and virtues should be part of the upbringing and education of children and young people to increase the likelihood they live well together in a diverse society. My illustrations come from the Netherlands, because I know this country the best. The second section explicates the term (super-)diversity and elucidates the central concept of identity. In the third section, I illustrate how diversity in the valuation of identity aspects, including valued character traits, is fostered by children's membership of families and communities. The fourth section describes two distinct ways in which societies respond to increased diversity. Adopting the response that accepts the fact of diversity, I suggest that citizens of such societies need to have a basic disposition that I call the double gaze, which is explicated in the fifth section. In the final section, I turn my focus to schools to describe their role in educating the double gaze.

Super-diversity of Western societies

Most Western European societies have changed in the way my village did. Particularly the big cities are now inhabited by citizens with a great variety in ethnic, national, and religious backgrounds. To express this, sociologists have begun to typify Western European countries as super-diverse. Steven Vertovec (2007) coined the term to draw attention to the fact that there is not only an increase in the diversity of ethnic groups that have immigrated into Western European countries (i.e., no longer only those from former colonies), but that ethnicity is but one aspect of diversity – migrants also have:

> differential legal statuses. . . , divergent labour market experiences, discrete configurations of gender and age, patterns of spatial distribution, and mixed local area responses by service providers and residents. The dynamic interaction of these variables is what is meant by super-diversity.
>
> (p. 1025)

While Vertovec used the term for changes in migration patterns, the term appealed to researchers from various disciplines and policy makers who gave it different and wider meanings (see Vertovec, 2019). I use it in a wider sense, too, namely that Western European countries are inhabited by people who differ in various respects, not only in ethnic and cultural background, but also with respect to their gender, sexual orientation, intelligence, social status, etc., and that they have different conceptions of the good. To call Western European countries (super-)diverse does not only give a more precise account of state of affairs, it also has the advantage that it does not exclusively focus on culture, ethnicity, or religion. This is advantageous, because the term multi-cultural has become rather blemished (e.g., Modood, 2016). More importantly, it does not make the mistake that people can be identified by one aspect of who they are (e.g., Appiah, 2005; Fukuyama, 2018; Sen, 2006; Young, 2000); a person's identity has various aspects.

Amélie Oksenberg Rorty and David Wong (1993) have made a helpful distinction between five clusters of aspects of identity: (1) somatic, proprioceptive, and kinaesthetic dispositions, e.g., being tall or having red hair; (2) central temperamental or psychological traits, e.g., being optimistic or agreeable; (3) social role identity, e.g., being a parent, a teacher, or an academic; (4) socially defined group identity, e.g., being a man, Dutch, or a Muslim; and (5) ideal identity, e.g., the aspiration to be a virtuous person. Of these clusters of aspects, ideal identity and socially defined group identity, comprising aspired and ascribed virtues, are particularly relevant for reflecting on (educating) local and universal virtues in a diverse society. As the meaning of ideal identity is clear, I only briefly explain socially defined group identity. This cluster is constituted through belonging to groups into which someone is born or of which one becomes a member in later life and to which (local) characteristics are ascribed and of which specific virtues are expected. Socially defined groups are, for instance, gender, race, ethnicity, class, and religion, but also a family or community. Yet, because traits are attributed to groups and not to individuals, they are not necessarily recognised by each individual and there may therefore be a gap between the way others characterise a person and the way in which a person typifies herself. Particularly when socially defined group identity is absolutised by others, and the person does not recognise herself in this representation, or when a person is no longer treated as an individual but as a representative of a group, social interaction is hindered and the well-being of a person may be seriously hampered (see also Blum, 2001).

People have many features, but not all of them are considered to be part of a person's identity. Which aspects are regarded as being significant is influenced by (cultural) values, for which Charles Taylor (1992) uses the metaphor of the horizon. For instance, the number of hairs does not tend to be included in a person's description of herself (Taylor, 1992, p. 36), while the type of a person's hair or the colour of her skin are regarded as relevant in Western European countries. These values also influence whether the aspect of the person is evaluated positively or negatively. For instance, women can cherish the empathic qualities that are commonly ascribed to them, religious believers tend to be proud of their faith and aspire to live up to their religion's ideals and virtues, while empathy can also be seen as a weakness and faith and devoutness as signs of backwardness (as, for instance, right-wing politicians in the Netherlands say about Muslims).[1] Thus, the way in which one describes oneself, including one's ideal identity, and constructs the identity of others, is a combination of personal aspects and one's values that influence one's beliefs of which characteristics are relevant, the evaluation of these characteristics, and one's beliefs toward which characteristics (including virtues) people should aspire.

The positive aspect of calling society super-diverse is that it puts less focus on social group identity (which of course does not mean that its citizens will also do so). However, this characterisation has the same disadvantage as the other terms; namely, that it emphasises the differences between people rather than their similarities. Yet, while human beings are unique persons, given the particular constellation of their characteristics and the values they endorse, they are also similar human beings. A profound similarity is that they have the same basic needs. Human beings have the same basic biological or physiological needs, i.e., the need for food, shelter, and security. They also have the same psychological needs. The self-determination theory of Richard Ryan and Edward Deci (e.g., 2000), for instance, argues and has proven with empirical research around the globe that all human beings have the need for autonomy, competence, and relatedness (which Baumeister and Leary [1995] call the need to belong). Other scientists have argued that human beings also have the psychological need to have meaning in life (e.g., Frankl, 1985; Baumeister, 1991) and the need for recognition (Honneth, 2001, 2004). For sure, the way in which people satisfy their needs and what they regard as being a virtuous way of doing so is diverse and influenced by the family and community in which they are raised and grow up, but recognising these similarities has implications for the way in which people interact with each other and for education.

Influences on identity formation and evaluation of the identity of others

Children construct the identity of themselves and others in interaction with others: they learn who they are through the reaction of parents, peers, teachers and other adults, and when they grow up, they begin to actively construct their identity by identifying with or distinguishing themselves from others. This includes not only (the evaluation of) the aspects that they have, but also the type of person they aspire to be, i.e., the (local or universal) virtues they value. I will illustrate this by the ways in which two environments contribute to identity construction; namely, the family and a community.

In raising their children, parents actively contribute to the identity development of children. Parents (aim to) influence all five identity aspects by the way in which they live with their children and the values they pass on to their children. For instance, from the remarks of their parents, children learn that they are kind, intelligent, or lazy, and with compliments and punishments, parents aim to foster a particular ideal identity – they want their child to be kind, use its intelligence to get high grades, and/or be a hard worker. Moreover, families have implicit and explicit rules and customs, beliefs, and values that regulate family life – what they do, what is allowed, and what is expected. In other words, families have a common culture in which certain characteristics are valued and in which particular virtues are fostered. The way in which the family perceives itself – i.e., the collective identity of the family – is also part of the identity of children. Parents tell their children what is characteristic of their family – e.g., that members of this family do not eat meat or that this family is proud to be Dutch – and children pick this up through stories of their parents; for instance, that the family suffered much adversity in their home country but has been able to cope and survive the migration to another country by being strong and hopeful.

Parents will underscore the particularities of both the individual identity of children, as well as the collective family identity, but these can phrased more or less oppositional to the identity of other people and families. And this also influences the way in which children learn to position themselves against others (e.g., Haste, 2004) – parents shape the way in which children perceive of or construct the identity of "others". For example, I was raised in a relatively orthodox Protestant family. Being a Protestant meant having particular virtues; for instance, being reliable, punctual, and frugal. We also picked up that these were not as typical of Catholics, who were perceived as being more lenient about their principles and flexible in agreements.

These influences are also found in communities of which children are members (via their parents). As, communities also have a common culture that "contains conceptions of character, of what a good person is like, and of the virtues that define such character" (Bellah, Madsen, Sullivan, Swidler, & Tipton, 1996, p. 153). These conceptions of character are related to the particular collective identity of communities, consisting of core shared values, norms and meaning (Etzioni, 2015; see also Bellah et al., 1996), which communities expect to be(come) part of the identity of its members as well. Similar to families, the collective identity of communities can be more or less defined in contrast with other communities. Iris Marion Young, for instance, notes that communities that ascribe themselves a particular identity "generate a logic of hierarchical opposition. . . . Any definition or category creates an inside/outside distinction, and the logic of identity seeks to keep those borders firmly drawn" (1986, p. 3). Such, what Nel Noddings (1996) calls normocentricity can lead to both admirable and deplorable results for both members of the community and outsiders. Liberal religious communities, for instance, tend to be open to internal diversity and have an open view to the outside world, even though they ascribe a clear identity to themselves. However, communities can also have dark sides like its tendencies toward parochialism, conformity, exclusion, and distrust (or hatred) of outsiders (Noddings, 1996, p. 258). Belonging to such a community means that children learn not only that they have to be similar to the members of their community (e.g., Young, 1986, p. 7), but also that their differences from others are important.

Thus, families and communities influence (ideal) identity by their local common culture and collective identity, and this contributes to diversity in society, as well as the way in which children interact with others. When differences are emphasised and when families and communities are closed, they diminish the likelihood of peaceful cohabitation in society. In the words of Robert Putnam (2007), families and communities may have a lot of bonding social capital, but may be low in bridging social capital that is needed to live well together with others. The bonding capital is, however, valuable for its members – families and communities are importance sources for satisfying the need of belonging. Although there are certainly exceptions, human beings are deeply attached to their families and communities, and are hurt when these are treated unfairly. For most members, their allegiances (also to the values and culture of the family and community) are a significant contribution to their sense of themselves. Thus, seeing others as similar human beings also encloses recognising the importance of these groups for them.

This is something else than recognising or accepting the groups' values, to which I will return later in this chapter.

Responses of societies and citizens to super-diversity

As mentioned before, the rise in diversity has led to an increase in tensions between the inhabitants of European societies. One of the main concerns is that citizens do not sufficiently share a collective identity supported by a common culture. A common culture can be seen as a glue of society: sharing a common culture supports social cohesion and strengthens citizens' allegiance to a society – a willingness to contribute to society or at least not undermine it, and a commitment to stand up for society in times of adversity (e.g., Raz, 1993). I believe that for these reasons, a common culture is important for society. Yet, there are different types of common culture. I will briefly describe and evaluate two types that deal differently with diversity between persons and similarities between human beings, realising that there are many variants in between.

The first way in which a society can deal with super-diversity is by aiming to diminish the influence of the diversity of collective and personal identities at the societal level by ensuring that all citizens share a substantive common culture that is characteristic of a nation. This route underscores the social group identity of the citizens of a country; it is local and different from other nations. We see this in modest forms in most European countries in the promotion of Dutch, British, French, or Danish values. In the Netherlands, not only the newcomers, but the entire younger generation, is to be immersed in Dutch history and customs to counter perceived lack of social cohesion and allegiance of citizens to the Netherlands. It is, for instance, compulsory for schools to organise an excursion to the Rijksmuseum.

Excessive nostalgic nationalism can be rejected on reasons of discrimination and the danger of a centralistic and near totalitarian state form that is necessary to maintain the uniformity of the political and public spheres, which will also blur the boundaries between the public and private sphere. My pedagogical or educative objection is that nostalgic nationalism diminishes the ability and possibility of the new, diverse generation of citizens to develop their own common culture. The world is changing continuously due to, for instance, technological innovations and migration, which means that lives of people change, as well. In my view, the current adult generation should therefore pass on values and cultivate virtues with which the new generation can form a society that is able to deal with new circumstances, new ideas, and ideals that are formed by debates between diverse groups (which does not necessarily mean that they will come to agreements about all aspects of society), instead of aiming to determine the future with specific historical cultural beliefs.

The second type of common culture respects diversity, in the public and political domain, too, but citizens are also expected to share the values and customs that are necessary to interact and cooperate with people who have different identities and conceptions of the good. Such a common culture is pursued in liberal democratic societies.[2] It consists of moral values or principles and civic virtues that are important for every decent society such as solidarity and justice, as well those that are characteristic of a liberal democracy, such as freedom and tolerance.[3]Citizens are expected (and therefore educated) to be intrinsically motivated to respect the laws of the liberal democracy; be tolerant of the ideas, practices, and customs of others; respect the equal rights of citizens; and be willing to contribute to the possibility that all citizens can live a minimally decent life. As a democracy, citizens are able to actively contribute to the politics and governments of the country, and they are expected to do so in a conscientious, critically reflective manner. This common culture is formal rather than substantive, has universal rather than local aspirations, and has a moral-civic character, i.e., it comprises the way in which people are expected to interact in the civil society and deal with political and public matters.[4]

I favour this second common culture, as citizens benefit from living in a society where they have the freedom to live their lives from the inside in families and communities, where they are treated equally as citizens irrespective of their identity, and where they can influence the policy of local and national governments. Maintaining a proper balance between freedom, equality, and solidarity in a society with diverse ideas about the interpretation of these principles and in ever-changing circumstances of the world is, however, challenging. A disposition that I want to suggest is vital for the interaction between citizens is what I call the double gaze.

The double gaze

As I have hinted at several points, in my view, people have to see the complete identity of other persons, as well as the similarity between themselves and other human beings. I call this disposition the double gaze. I qualify it as a disposition rather than a virtue, because it is not a mean. While one can be lacking, there is no excess. If one sees the different aspects or similarity of the other only, one simply does not look with a double gaze. Furthermore, it is less context-dependent than a virtue, as it sets similar standards for all people in their encounter with everyone else. The idea of the double gaze is inspired by Iris Murdoch's concept of attention. Attention, says Murdoch, is "a just and loving gaze directed upon an individual reality" (1970, p. 34; see also Blum, 1991; Cordner, 2016). I rephrase this as the intention of people to look to others with care; to carefully see others as human beings[5] who have the same basic needs and as persons with various characteristics who are also member of a family and community with values to which they might be deeply attached. The double gaze is thus both blind to difference and a precise vision of the richness of the identity of the other.[6] For sure, we cannot immediately see all the details, and therefore, it should be phrased more modestly that one should look beyond one's first impression.

The double gaze does not imply that people have to accept all the aspects that others believe to be characteristic of themselves. When we educate children, this is quite self-evident – parents and teachers aim to influence characteristics that they regard to be undesirable (whether or not they are right is another matter). For instance, short-tempered children learn that they should contain their anger and express their feelings of discontent mildly; in traditional families, girls are taught that it is characteristic of women to obey their husbands, and that they should therefore tone down their ambitions. This is, however, also true for the interaction between adults. There is a difference between recognising that someone is a human being and a person, and valuing the traits or the values and ideals of a person. Citizens may reject each other's lifestyles, but they have to want to encounter them as human beings and as persons with possibly deep-rooted ideas to which they attach high value. Equally, they do not have to be kind to everyone; behaviour of others that one regards as unacceptable does not have to be condoned. Yet, when one, for instance, tells a group of youngsters who are harassing an elderly person that they should treat elderly people with respect, one does not approach them as migrants, as (spoiled) rich kids or "those kids from the deprived neighbourhoods", but as individuals and as human beings and wonders what drives them to behave in this manner. Such a disposition is not only more careful, I also suspect (but I have no empirical evidence for this, only personal experience) that this will be more effective as each young person is approached carefully and fairly and may therefore be more willing to listen.

The double gaze is not easy. People quickly label others, also on the basis of their implicit associations that are informed by their values and experiences. While this might be a common and even helpful quality (to respond quickly in unknown situations), it also diminishes the possibility of an open encounter, particularly when one is informed by stereotypes. Although I suggest that the double gaze is a universal disposition, it is a particularly vital aspect of the virtues to live well in a super-diverse society and should therefore inform the theory about and practice of desirable virtues and their education in such a society. I give three examples.[7]

First, it can be regarded as an aspect of the virtues of tolerance and open-mindedness. These, according to Howard Curzer, consist of "dispositions for being appropriately open to the difference of valuation expressed in novel actions, fresh ideas and alternative ways of life" (2012, p. 81). Characteristic for the double gaze is that people look beyond their first impression. In looking at and interacting with people in the proposed double manner, one is primarily curious instead of making snap judgements on the basis of the visible aspects of a person. Moreover, in interacting with the other, one wonders if one should give primacy to a particular aspect of a person's identity (and which aspect that would be), or if one should treat the other as a human being and therefore similar to other human beings, irrespective of their specific characteristics.

Second, there is an iterative relation between looking at others with an open view and placing one's own local values and ideas in the background and the virtue of humility. As an epistemic virtue, similar to openness, it means that one accepts the possibility that one might be mistaken about others, and that one's own beliefs could be wrong. The double gaze means that one does not position oneself above the other, but next to the other. Neither does one position oneself under the other, for accepting all aspects of the other or her convictions without critical reflection would be a form of blind submission.

Finally, the double gaze is conducive to (and an aspect of) the virtue of civility. I understand civility in two senses or two types of virtue (see also Curzer, 2012). First, civility is a quality of communication between persons and groups, as Cheshire Calhoun understands it (2000). According to Calhoun, civility is a distinct moral virtue, namely the way in which people communicate with others or display a moral virtue like respect or tolerance. It aims to "safeguard the possibility of a common social life together" (p. 272), which may conflict with one's own moral judgement about the convictions of the other. The double gaze draws one's attention to the importance of engaging with others in a civilised manner to continue dialogue rather than ignoring or vilifying others. Second, civility can be regarded as a political virtue or duty, as John Rawls, for instance, suggests. According to Rawls: "The ideal of citizenship involves a willingness to listen to others and a fair-mindedness in deciding when accommodations to their views should reasonably be made" (1993, p. 217). And to be able to do this in a diverse society requires the disposition of the double gaze.

The role of schools in fostering the double gaze

The exposé about super-diversity of society and commonality of human beings shows that locality can be used in two different senses. First, locality can be used for the group or person having the values or common culture, e.g., families, communities or societies. Second, it can be used for the extension or the realm of the people who are the subject of one's action. Values and virtues can be relevant in interaction with a particular person (e.g., one's sibling or teacher) or within a group (e.g., family or school), and there are values and virtues regarding one's interaction with all people. And the way in which one interacts with unknown others can in turn be particular (i.e., that the characteristics of the other affect the way in which one acts), or be similar (i.e., that one's action is impartial or person-indifferent), or both (e.g., that one helps another person who is seriously hurt, but one helps everyone who is hurt irrespective of their appearances). These distinctions are normally part of the education of children. Yet in a diverse society there are different ideas about these "rules of interaction, cooperation and cohabitation", and the ones that children learn in families and communities can be more or less conducive to living well in a diverse society.

As states and citizens cannot be certain that children learn that the moral virtues of honesty, compassion, humility, modesty, and justice are relevant for one's interaction with all other human beings and that all members of society have equal rights as citizens, and because families and communities are regarded as private spaces, schools are given responsibility for the moral and citizenship education of the next generation, for in schools, the state can exert its influence in both the children's and society's interests. Schools are in theory well placed to do so, for in school children from families with various backgrounds and communities with different values have to interact and work together. The local common cultures are brought into the school. For sure, pupils have more opportunity to practice the double gaze in schools consisting of pupils with different religious, ethnic, and national backgrounds, but in schools that are relatively homogeneous in those respects, students also have different identities and come from a diversity of families and from various communities.[8]

Schools have their own common cultures related to their history and denomination, formed by the teachers, school leaders, students, and parents, and influenced by (their relation to) the immediate and wider environment. Yet, this common culture does have boundaries, and these are the boundaries of the liberal democracy; positively framed, it should be a culture in which the similarity and diversity of students is recognised and whereby differences in identity are treated in a respectful manner, even when they are not valued. This should be reflected in what students learn, the disposition of teachers, and the ethos of the school. I will illustrate this with a few examples.

Understanding that other people are human beings who have the same basic needs and similar vulnerabilities is not difficult. Yet, treating others like one wants to be treated as a human being is not always easy; students are therefore reminded of the Golden Rule to do so. Teachers also have ample opportunity to draw attention to human similarities in various curricular subjects, like geography, history, and citizenship education, and by investigating human rights with students, but also by reading books and watching films. Yet, the Golden Rule and the curricular subjects are also used to promote students' understanding that people differ from each other and therefore should be treated differently – and precisely dealing with this flipping of perspectives is valuable for students to practice.

Grasping the complete identity of other persons is much more difficult, because of the various aspects it consists of and the variety in evaluation of identity aspects. Learning the disposition of the double gaze (therefore) also requires knowledge of conceptions of the good that inform these evaluations. I suggest that particular attention be given to the (history of) local cultures of families and communities of children present in the school and the highly prevalent local common cultures in society. This not only assists students to better understand their fellow students, but also contributes to the satisfaction of the need of pupils to be recognised that they belong to families and communities to which they are attached. Moreover, students learn to see that there is diversity within those groups and that they consist of individuals who each have other characteristics as well (see also Blum, 2001). Thus, in the Netherlands, pupils should learn about Dutch culture and its history and main religions, but also those of Turkey and Morocco, as these form the main groups of migrants, and of people from the former colonies (Indonesia, Surinam, and the Dutch Antilles). In this respect the heated and sometimes violent debate about the Dutch tradition of Zwarte Piet (Black Pete), who accompanies St. Nicolas on his visit to the Netherlands each year, which is defended by some Dutch groups and seen as highly denigrating by people of colour, is an important topic to discuss in upper levels of primary schools. Another heated societal debate particularly suitable for secondary schools is the description of and perspective on Dutch (colonial) history. Finally, the way in which the media reports or citizens and politicians talk about migrants, Muslims, women, LGBT, but also the poor as being a seemingly homogeneous group, is an endless source for reflection in the classroom. Teachers should not aim to promote a particular position, nor to convince students of a negative evaluation of a position unless it clearly conflicts with liberal democratic values (and this is often a point of contention), for this would impose a controversial point of view. Their aim should be that students see others better and think about ideas and values from different perspectives.

Good teachers exemplify the double gaze in their classes all the time, and by explaining to students (now and then) why they act in a particular manner, students gradually learn what the double gaze involves. Sensitivity of teachers is not only conducive to satisfying individual students' needs of belonging and recognition, but also facilitates to seize moments – Kairos moments, as Max van Manen (2016) calls them – to enhance students' careful encounter with each other and the manner in which they talk about other people. This may cause unease and opposition, because it does confront students with their own values and ideas. Yet, a teacher who has pedagogical tact (also a concept of Van Manen) is able to prevent students from being publically humiliated or vicious tensions between groups of students arising, but to realise that these moments lead to reflection and a willingness to be open and civil even if one continues to have different ideas. This, in turn, requires a safe school environment.

The ethos of the school is essential for cultivating the double gaze. A newspaper article on discrimination and racism in schools in Amsterdam (*Het Parool*, 29 December 2018) illuminated how common it is for secondary school students in both "white" and "diverse" schools to use negative stereotypical labels when addressing other pupils. Teachers and school leaders do intervene, but try not to be moralising, as they believe this will be counterproductive. While I cannot judge the best approach, as this will be different in every school, teachers do have to be tenacious in their pursuit, for the ethos of the school can foster students to recognise that there is more to their fellow students and teachers than what is immediately visible, just as it can encourage students to fully expose themselves to others. This provides an excellent space to practice the double gaze that they need outside the school.

In conclusion

My contribution has its limitations. I have only touched upon one aspect of moral and citizenship education, and I have hardly dealt with political issues in and around schools and in society at large. Furthermore, theorists who defend a nationalistic or religious idea of society may believe that I have not sufficiently addressed their ideas. For others, my proposal may not go far enough, as it does not defend that in principle, the values and common culture of persons, families, and communities should be recognised as being valuable. Teachers, finally, may argue that things are well said but that I underestimate the challenges they face in classes of 30 students and a full curriculum. Yet, I hope that my focus on a disposition that I believe is vital for living well together in a (super-)diverse society and describing ways in which educational spheres can encourage and inhibit the development of the double gaze has given food for further philosophical and educational reflection. With respect to the basic question motivating the present volume, my analysis of the double gaze has aimed to show how a universal disposition should affect the local and universal virtues taught and caught in a super-diverse society.

Notes

1 Thus, social group identity can be a source of pride and the formation of movements that strive for the recognition of their identity and positive discrimination and of shame, as well as a ground of recognition and discrimination by "outsiders" (e.g., Rosenblum, 1994; Young, 2000).

2 See Raz (1993) for a proposal of a liberal democratic common culture of a diverse society.

3 See Kristjánsson (2002, 2006) for an extensive discussion about the compatibility and hierarchy of these two dimensions. See also Arthur, Kristjánsson, Harrison, Sanderse, and Wright (2017).

4 It might be challenged that this can be called a common culture. Appiah (1996), for instance, uses a more "demanding" interpretation, namely that it is characteristic for people sharing a common culture that this culture centres their lives and identity. A thin political common culture is arguably not substantive enough to have this function. Furthermore, he suggests that it is not necessary that all citizens are committed to societies' values in the same way, or for the same reasons. I agree, but this is a strict interpretation of common culture, which also makes it difficult to say that communities have a common culture. I therefore use it in a looser sense.

5 Appiah (1996) quotes de Maistre, who has learned from Montesquieu that he has never met man, only seen people with a particular identity. Appiah himself

believes that "we can treat others decently, humanely, *through* our differences. The humanist requires to put our differences aside; the cosmopolitan insists that sometimes it is the differences we bring to the table that make it rewarding to interact at all" (p. 25). In my view, both are important.

6 The double gaze also clarifies that equality should be understood in two different senses; namely as similarity (of human beings) and as having equal rights as citizens who have diverse identities.

7 There is no space to deal with other virtues, but I hope my explication of the double gaze is sufficiently clear to realize that it is both an aspect of compassion and justice and enabling to be compassionate and just.

8 For a pessimistic view, see Merry (2018).

References

Appiah, K. A (1996). Against national culture. *English in Africa*, *23*(1), 1–27.

Appiah, K. A. (2005). *The ethics of identity*. Princeton: Princeton University Press.

Arthur, J., Kristjánsson, K., Harrison, T., Sanderse, W., & Wright, D. (2017). *Teaching character and virtue in schools*. London: Routledge.

Baumeister, R. F. (1991). *Meanings of life*. New York: The Guildford Press.

Baumeister, R. F., & Leary, M. R. (1995). The need to belong: Desire for interpersonal attachments as a fundamental human motivation. *Psychological Bulletin*, *117*, 497–529.

Bellah, R., Madsen, R., Sullivan, W., Swidler, A., & Tipton, S. M. (1996). *Habits of the heart: Individualism and commitment in American life*. Updated edition with a new introduction. Berkley: University of California Press.

Blum, L. (1991). Moral perception and particularity. *Ethics*, *101*(4), 701–725.

Blum, L. (2001). Recognition and multiculturalism in education. *Journal of Philosophy of Education*, *35*(4), 539–559.

Calhoun, C. (2000). The virtue of civility. *Philosophy and Public Affairs*, *29*(3), 251–275.

Cordner, C. (2016). Lessons of Murdochian attention. *Sophia*, *55*, 197–213.

Curzer, H. J. (2012). An Aristotelian account of civility. In D. S. Mower & W. L. Robinson (Eds.), *Civility in politics and education* (pp. 80–98). London: Routledge.

Etzioni, A. (2015). Common good. In M. T. Gibbons (Ed.), *The encyclopedia of political thought*. John Wiley & Sons, Ltd. Retrieved from https://icps.gwu.edu/sites/g/files/zaxdzs1736/f/downloads/Common%20Good.Etzioni.pdf

Frankl, V. E. (1985). *Man's search for meaning*. New York: Washington Square Press. (Original in German, 1946).

Fukuyama, F. (2018). *Identity: Contemporary identity politics and the struggle for recognition*. London: Profile books.

Haste, H. (2004). Constructing the citizen. *Political Psychology*, *25*, 413–439.

Honneth, A. (2001). Recognition or redistribution? Changing perspectives on the moral order of society. *Theory, Culture and Society*, *18*(22–23), 43–55.

Honneth, A. (2004). Recognition and justice: Outline of a plural theory of justice. *Acta Sociologica*, *47*(4), 351–364.

Kristjánsson, K. (2002). In defence of "non – expansive" character education. *Journal of Philosophy of Education*, *36*(2), 135–156.

Kristjánsson, K. (2006). *Justice and desert-based emotions*. London: Routledge.

van Manen, M. (2016). *Pedagogical tact: Knowing what to do when you don't know what to do*. London: Routledge.

Merry, M. S. (2018). Can schools teach citizenship? *Discourse: Studies in the Cultural Politics of Education*, *41*(1), 124–138. https://doi.org/10.1080/01596306.2018.1488242.

Modood, T. (2016). Multiculturalism. In G. Ritzer (Ed.), *The Blackwell encyclopedia of sociology*. https://doi.org/10.1002/9781405165518

Murdoch, I. (1970). *The sovereignty of good*. London: Routledge.

Noddings, N. (1996). On community. *Educational Theory*, *46*(3), 245–267.

Oksenberg Rorty, A., & Wong, D. (1993). Aspects of identity and agency. In O. Flanagan & A. Oksenberg Rorty (Eds.), *Identity, character and morality: Essays in moral psychology* (pp. 19–36). Cambridge: MIT Press.

Putnam, R. D. (2007). *E Pluribus Unum*: Diversity and community in the twenty-first century. The 2006 Johan Skytte prize lecture. *Scandinavian Political Studies*, *30*, 137–174.

Rawls, J. (1993). *Political liberalism*. New York: Columbia University Press.

Raz, J. (1993). *Multiculturalism: A liberal perspective*. Retrieved from malikdolgozat.uw.hu/cikk/Raz_Multiculturalism.pdf

Rosenblum, N. L. (1994). Democratic character and community. The logic of congruence? *The Journal of Political Philosophy*, *2*(1), 67–97.

Ryan, R. M., & Deci, E. L. (2000). Self-determination theory and the facilitation of intrinsic motivation, social development and well-being. *American Psychologist*, *55*(1), 68–78.

Sen, A. (2006). *Identity and violence: The illusion of destiny*. New York and London: W.W. Norton & Company Ltd.

Taylor, C. (1992). *The ethics of authenticity*. Cambridge: Harvard University Press.

Vertovec, S. (2007). Super-diversity and its implications. *Ethnic and Racial Studies*, *30*(6), 1024–1054.

Vertovec, S. (2019). Talking around super-diversity. *Ethnic and Racial Studies*, *42*(1), 125–139.

Young, I. M. (1986). The ideal of community and the politics of difference. *Social Theory and Practice*, *12*(1), 1–26.

Young, I. M. (2000). *Inclusion and democracy*. Oxford: Oxford University Press.

12 Between the local and global

Placing the virtuous professional practitioner in context

Sarah Banks

Introduction

This chapter explores the local–global dynamics of moral virtues with reference to narratives from social workers around the world about ethical challenges in practice. Social work provides a particularly fruitful focus for the debate about local versus global virtues, as it can be construed as a locally based practice as well as a nationally and internationally recognised profession and social movement. There is also growing concern about professional and ethical "imperialism" (exporting inappropriate theories, values, and practices from the global North to the global South), alongside increasing interest in the potential of virtue ethics as a more widely recognised discourse than the principle-based language of individual rights and utility that predominates in international statements on ethics. The chapter considers whether virtue-based ethics is more universally applicable, given that virtue-based language/concepts are commonplace in the global South, and similar character traits may be valued as important across cultures and nations, e.g., trustworthiness, compassion, and honesty. After examining the most recent principle-based international statements on ethics in social work, the chapter considers two case examples of ethical challenges from practitioners in different parts of the world (Palestine and Vietnam), including how they construct the "ethically good social worker". The value of working with cases is discussed, including the importance of hermeneutical dialogue between reader and the case, attention to the details of the context of the case, and a commitment to undertaking the hard work of transcultural ethical engagement.

In this chapter, I take a moral virtue to be a disposition to feel, think, and act in such a way as to promote human and ecological flourishing, entailing both a motivation to act well and, typically, the achievement of good ends. Virtues are often described as excellent traits of character and entail a reliable disposition to act in certain predictable ways across contexts.

Social work as local, national, and global

As a local practice, frequently regarded as part of an occupation or profession at a national level, social work is concerned with enabling and empowering people to function better in their local contexts (families, neighbourhoods, and institutions) and/or to change features of their lives and environments. Social work is usually part of a welfare system, and may be directly or indirectly funded and sanctioned by local or national states. Hence, the role of social work practitioners, and what is regarded as effective and ethical practice, is in important respects relative to particular circumstances, relationships, resources, and welfare and political regimes. Indeed, what is called "social work" in one country may be significantly different from "social work" in another country. For example, in the United States, a significant amount of social work happens in private practice, with an emphasis on "clinical" work (casework and therapy). In South Africa, much social work is provided by non-governmental organisations, with a focus on social development (including neighbourhood-based group work and community development).

Yet, in spite of local variations, as with many occupations and professions, there are international associations for social work, which provide opportunities for networking, sharing good practice, lobbying, and policy advocacy on pressing social issues. The most significant of these are the International Federation of Social Workers (IFSW) and International Association of Schools of Social Work (IASSW). These associations also play a role in developing and maintaining the status and identity of the profession, which, despite its many different manifestations across the world, shares a common concern in working with people experiencing social difficulties in their lives. To facilitate this common identity, there is an international definition of social work (IASSW & IFSW, 2014), global standards for social work education (IASSW & IFSW, 2004b), and a global statement of ethical principles (IASSW, 2018; IFSW, 2018). The current definition of social work is as follows:

> Social work is a practice-based profession and an academic discipline that promotes social change and development, social cohesion, and the empowerment and liberation of people. Principles of social justice, human rights, collective responsibility and respect for diversities are central to social work. Underpinned by theories of social work, social sciences, humanities and indigenous knowledge, social work engages people and structures to address life challenges and enhance wellbeing. The above definition may be amplified at national and/or regional levels.
>
> (IASSW & IFSW, 2014)

Not surprisingly, the drafting, agreement to, and revision of the definition and other documents have been, and remain, both challenging and controversial. The documents need to be pitched at a sufficiently generic level to be applicable to all members across the world. Yet at the same time, they must also be specific enough to outlaw practices agreed to be unacceptable and ensure certain valued principles or standards of practice are not misinterpreted or ignored, whilst also acknowledging national and cultural differences. In creating the most recent revised statement of ethical principles during 2017–2018, the drafters were concerned to challenge the dominance of western ethical perspectives based on individualism and the status of professionals as detached experts, balancing this with a more relational and democratic approach.

The recently revised "Global social work statement of ethical principles", eventually published in longer and shorter versions (see IASSW, 2018; IFSW, 2018), comprises general principles about respecting human dignity and promoting human rights and social justice, whilst also outlawing use of torture in social work, stressing respect for confidentiality of information relating to service users, promoting anti-discrimination, respecting diversity, and acknowledging that taking gifts may be acceptable in some countries. It is an amalgam of ethical and practical principles and some more specific rules, and seems to have lost some coherence in its latest re-drafting. This is not surprising, given significant disagreements over its content, resulting in the two associations (IFSW and IASSW) producing different versions of the statement. In this sense, the documentation reflects the current state of debate in the international social work associations, at a point when existing consensus has broken down and a new consensus has not yet emerged. Arguably, this situation is partly a response to the growing critique of the dominance of Western values and ethical principles in international social work, resulting in the superimposing of value statements of different types and levels of specificity onto the previous universal human rights-oriented document (IASSW & IFSW, 2004a).

The IASSW contribution to the draft statement resulted in a document that was quite theoretically based, referring to concepts of "embodied vulnerability" (shared between social workers and service users), the "moral impulse" of social workers, recognition of inter-subjectivity, and human beings as embedded in societies (rather than autonomous individual agents). A footnote (IASSW, 2018, p. 1) refers to "western hegemony" and a "decolonising agenda". Since IFSW did not feel able to endorse the inaccessible language and academic orientation of parts of the document, it was agreed that it would publish a simpler and shorter version (IFSW, 2018). The ethical principles and action guidance are broadly the same, but the IASSW document also articulates a relational worldview in parallel to the principle-based approach.

It may be helpful to understand this situation in the context of the development of social work, which first became established as an occupation in Europe and North America in the mid- to late nineteenth century, developing from voluntary and charitable social welfare work to support people in need, control social problems, and organise for social change. Social work is now established in most countries, very much influenced by theories, models, and practices from the global North. This has led to accusations of "professional imperialism" (Midgley, 1981); that is, the exporting of inappropriate models of professional social work based, for example, on individual casework and psychoanalytic theories. These have been transferred to countries and cultures where the predominant forms of social organisation may comprise extended families, tribes, and other collectivities and communities, and/or the most pressing social issues are extreme poverty, environmental degradation, or humanitarian crises, which call for social development rather than casework (Chitereka, 2009; Hugman, 2010, 2013; Midgley, 2017). On this view, it is important that "indigenous" or "culturally relevant" forms of social work develop that fit local contexts, needs, and priorities (Gray, Coates, & Yellow Bird, 2008; Gray, Coates, Yellow Bird, & Hetherington, 2013; Osei-Hwedie, 1993), and also that statements on ethics, textbooks, and models of practice reflect indigenous values and ways of working (Banks, 2015; Hugman, 2013; Weaver, 2016).

Universality of virtue discourse in everyday life

It has been suggested that looking at ethics in professional life in terms of good qualities of character or virtues may be more universally acceptable than the principle-based, individual rights-oriented approach that is currently dominant (see Banks & Gallagher, 2009; Banks, 2012). According to Gyekye (2010), character forms the basis of African ethics. Indeed, the language of virtue appears in many religious texts, from Confucius to Aquinas. These texts may not always concur on which virtues are most important, or how a particular virtue should be interpreted, but the language of good character or virtue (and vice) is present. Indeed, when Dahlsgaard, Petersen, and Seligman (2005) studied written texts from influential traditions of thought in world history (Confucianism, Taoism, Hinduism, Athenian philosophy, Judaism, Christianity, and Islam), they found "a surprising amount of similarity". However, while their literature review converged on six core virtues – courage, justice, humanity, temperance, wisdom, and transcendence (Dahlsgaard et al., 2005, pp. 204–205)[1] – it is important to note that all six virtues were not present in all traditions (only justice and humanity were explicitly named in all). The researchers also found variability across cultures in what was most esteemed, and many virtues were discounted because they were not ubiquitous – that is, they were culture-bound.

This echoes a recent empirical study conducted amongst students from 14 nations (from Malaysia to the Netherlands), which found core virtues shared by more than half the nations (van Oudenhoven et al., 2014). These virtues were: honesty, respect, kindness, openness, and tolerance, based on participants' answers to the open-ended question: "What do you find important personal characteristics which you would like to bring into practice in daily life?" It is not clear what further information participants were given regarding what would count as an "important personal characteristic", since this is not the same as the definition of virtues in the paper as "morally good traits that everyone either may possess or can learn" (p. 2).

Participants were also asked in a closed question to rate a set of virtues gathered from a previous study in the Netherlands. The authors conclude that there are strong differences between nations in terms of importance ratings, and in the frequencies of virtues given in response to the open-ended question. For example, while honesty was mentioned in all nations, generosity was only mentioned by French respondents. Similar nations with respect to language, culture, and/or religion also tend to share importance ratings. Regarding the relative universality of virtues, the authors refrain from drawing strong conclusions on the basis of data from 14 nations, but suggest the five most frequently mentioned virtues previously listed are potential candidates.

Such studies have many limitations, including ensuring translations between different languages capture nuances of meanings of virtue concepts, and difficulties of interpretation and comparability between countries and cultures. In the last study, respondents do not seem to have been asked what they meant by the virtues they listed, so we are left with a set of names of virtue concepts, but no description of what they mean. This is problematic, as virtue concepts are often characterised as "thick" (Williams, 1985, pp. 129–130; Annas, 2016). They have a detailed descriptive meaning relative to the contexts in which they are used, as well as an evaluative meaning. This contrasts with deontological, action-oriented concepts such as "right", "wrong", "ought", and "duty", which have an evaluative meaning, but no descriptive content inherent in the concepts themselves. For example, saying to someone, "It is right to break confidentiality here" gives no clue as to why it is right; whereas saying, "Be courageous and break confidentiality here" enables the addressee to work out what is entailed – perhaps taking a stand and risking adverse consequences. Although each person will have, or will develop, their own motives for, and understanding of, courageous action in a particular case, they will be able to find content in the injunction to be courageous. This is the kind of argument used by Williams (1985) and Annas (2016) for virtue concepts being regarded as "thick".

However, if we accept this characterisation of virtue concepts as essentially thick, then the lists of names of moral qualities derived from the international research mentioned earlier are relatively "thin" versions of virtue concepts. They do have some descriptive content, but it is relatively abstract and generic (for example, courage relates to making an effort to do something when some degree of risk is involved). The kind of survey research undertaken by van Oudenhoven et al. (2014) is even more problematic if we take a particular view of the evaluative element of thick concepts. For example, Väyrynen (2013, quoted in Cline, 2015, p. 10) argues that the evaluations conveyed by thick concepts are not part of their semantic meaning, but rather occur pragmatically from the shared evaluative backgrounds of speakers in a linguistic community. Arguably, this means that conducting questionnaire surveys about the importance and prevalence of virtue concepts in different countries gives us limited insight into the contextualised nuances of locally generated meanings that are in the very nature of virtue talk.

Hence, these potentially universal virtue concepts identified in the research are somewhat abstract. Arguably, we need to relocate them in their cultural contexts in order to engage in ethical dialogue between people from different nations and cultures, to rehearse moral disagreements and (sometimes) reach consensus on ethical evaluations. As Flanagan (2017) argues:

> Virtues that engender widespread agreement when described abstractly can conceal disagreement and regimens of oppression when one gets down to the nitty gritty details.
>
> (p. 6)

Flanagan's concept of moral ecology is useful here in its emphasis on ethics as concerned with human well-being in particular natural and social environments (p. 44). In the next section, we explore the moral ecology of professional life, drawing on contextualised accounts of social workers' lived experiences.

Virtues in professional life: insights from social work ethics cases

The two empirical studies previously mentioned relate to virtues in everyday life. They suggest that some virtues may be universally recognised and valued (albeit described in a fairly "thin" way). In professional life, these everyday virtues will also be relevant, but there are likely to be specific thicker virtues, or specific interpretations of virtues, that may be valued in professional contexts. While the literature on virtue ethics in social work is still relatively small, there is a growing interest (Banks, 2016b; Banks & Gallagher, 2009; Houston, 2003; McBeath & Webb, 2002; Pullen-Sansfacon, 2010; Wulfekühler & Rhodes, 2018), which draws on literature on virtue ethics in the professions more generally (Bondi, Carr, Clark, & Clegg, 2011; Oakley & Cocking, 2001; Carr, 2018). In order specifically to explore the moral virtues exercised in social work around the world, I now consider what we can learn from accounts written by social work practitioners of situations they found ethically challenging.

During 2009–2011, 26 real life "ethics cases" were collected from social workers for a book of cases and commentaries (Banks & Nøhr, 2012). These cases come from a range of countries, from Peru to Finland, each comprising two to three pages of rich description of circumstances, events, and actions construed by the authors to have ethical import. They vary in the extent to which the authors offer accounts of their emotions and reflections in or on action. Unlike short "textbook" cases, often written to highlight an ethical dilemma or crisis point, the ethical dimensions of these stories are embedded in "thick" descriptions of events (detailed, contextualised, and rich with meaning and interpretation; see Geertz, 1973). The ethical dimensions are only occasionally explicitly identified in reflective comments by the authors. The accounts tend to be action-focused – telling what happened, what the author and other key actors said and did. Hence, the language of virtue is not widely used. Even when the authors of cases explain their actions, virtue terms do not usually feature in these explanations. This is not surprising, since as Williams (1985, p. 10) comments, virtue terms themselves generally do not occur in the content of ethical deliberations, for example: "The benevolent or kind-hearted person does benevolent things, but does them under other descriptions, such as 'she needs it', 'it will cheer him up', 'it will stop the pain' ". To elaborate further, "a courageous person does not typically choose acts as being courageous . . . a modest person does not act under the title of modesty" (Williams, 1985, p. 10). According to Williams (1985, p. 11):

> The deliberations of people who are generous or brave, and also the deliberations of people who are trying to be more generous or braver, are different from the deliberations of those who are not like that, but the difference does not mainly lie in their thinking about themselves in terms of generosity or courage.

Hence, in examining the ethics cases, readers have to interpret the accounts of actions and motives in terms of virtues. The cross-cultural dimension lies in the provocation the cases bring to the readers from different cultural contexts, who can ask themselves the following questions. Would I wish to be the sort of person who acts like this? Would this action be virtuous in my context? What are the features of this context that help make sense of the situation depicted? It is essentially the dialogue between the reader and the case as written that is necessary to bring virtue concepts and virtue talk into being. When used in teaching and learning about ethics in professional life, the role of the teacher is to stimulate a virtue-focused discussion based on the largely descriptive accounts of events and actions. This could be described as an hermeneutical approach to ethics, seeking to develop understandings of situations, being open and sensitive to unexpected aspects and different people's perspectives, and questioning one's own preconceptions and frames of reference (Gadamer, 2004; Widdershoven, 2005).

I now summarise two cases from Banks and Nøhr (2012), inevitably losing some of the richness of the longer narratives, but giving a flavour of the ethical issues and the responses of the social workers involved. In the book, each case is accompanied by two commentaries, at least one of which is written by someone from a country other than the one in which the case happens. Names of service users have been changed.

Courage or foolhardiness?

Case example 1 Responding to child sexual abuse in East Jerusalem

This account comes from a school-based social worker in Palestine, in the Israeli-occupied territory of East Jerusalem. Here, school-based social workers are employed by the Palestinian Authority and United Nations Relief and Works Agency. However, the occupied territories are under Israeli law, which requires health and social care professionals, and members of the public, to report suspicions of child neglect and abuse to the Israeli authorities. Failure to report is a criminal offence.

Rana, a 14-year-old Palestinian girl, told her school social worker that her father was sexually abusing her. The social worker decided not to report this to the Israeli authorities as required by the law. Instead, the social worker and head teacher worked with the whole family to ensure the father lived in a separate part of the family house and Rana was able to lock her bedroom door. The reasons behind not reporting included the general reluctance to involve the Israeli authorities in the occupied territories, the likelihood of Rana and the whole family suffering shame and ostracism within their own community, the dangers of honour killing, and the poor provision for young women in the social services system.

In this case, the social worker and teacher were taking a risk in acting illegally and trusting that the father would change his behaviour and Rana would be kept safe. Considering this case in virtue terms, we could ask the question: were they acting courageously or in a foolhardy fashion? Although the social worker who contributed the case does not frame it in terms of virtue ethics, one of the commentaries on this case does:

> The school social worker appears to have been motivated by care for the young person. She was working, possibly instinctively, from a virtue ethics perspective, balancing rules and consequences and employing an understanding of what Rana needed to protect her immediately and to enable her to function well in her community and society in future.
>
> (Baidoun & Lindsey, 2012, p. 226)

Yet Baidoun and Lindsey also note that the father's behaviour is likely to be compulsive. This view is echoed in the second commentary, written from a

British perspective (Jones, 2012). While sympathetic to the circumstances, Jones (2012, pp. 230–231) cites Western research showing that abusers rarely desist and hence there is a likelihood that the abuse will continue and other young women may potentially be at risk. In both commentaries, the authors express reservations about the social worker's decision and actions. Yet as Baidoun and Lindsey comment: "Her actions were essentially contextually located and pragmatic" (2012, p. 227). We might suggest she was deploying the overarching virtue of *phronesis*, or professional ethical wisdom (Banks, 2018), based on her experience and knowledge of the context for social work in East Jerusalem.

Honesty or benevolence?

The second case (also taken from Banks and Nøhr, 2012, pp. 91–94) is given in more detail, as the author offers some reflections on her ethical position, with implications for the relativism/universalism debates. Here the author is a Dutch social work student working temporarily in a country very different from her own. This presents her with the challenge of how to evaluate (ethically and professionally) local beliefs and practices as an "insider-outsider" (she is present inside the situation, yet is an outsider culturally). She also gives her own reflexive commentary on her experiences and the issues raised for her. In this sense, through her writing, she is performing the role of the "ethically good" student – demonstrating herself doing the ethical work of being sensitive and respectful, and undertaking practical reasoning.

Case example 2 A Dutch student in Vietnam

A social work student from the Netherlands was undertaking a fieldwork practice placement in a rehabilitation hospital in Vietnam. She accompanied two physiotherapists on a home visit to an 8-year-old boy, Trung, who had Duchenne Muscular Dystrophy (DMD, a severe form of muscular dystrophy characterised by rapid muscle degeneration). Trung was in a wheelchair, writing on a small table attached to the chair. His mother proudly talked about the school prize he had won for his handwriting and Trung discussed his favourite activities. His mother asked when there would be an improvement in Trung's physical condition. One of the physiotherapists said he did not know, and that Trung should keep doing his exercises. On returning to the hospital, the student asked the staff if any improvement was possible with DMD. They replied "no". There was a possibility that Trung did not have a long life ahead. However, it was better that Trung and his parents did not know this; otherwise, they would feel sad until the time came for Trung.

The student reflected on this case:

> This approach is in contradiction with my social work ethics. But the staff explained to me that this is their way of working and it is better for the boy not to feel sad, because in their opinion that might increase the weakness of the muscles. Another argument for them not to tell the truth about the prognosis was the quality of the boy's life. According to the medical staff, if Trung knew that he would die, for example, in less than a year, he would spend the rest of his life miserable knowing that.
>
> From my point of view as a social worker, this could not be determined by the medical staff, for as far as we know, the boy could spend his last months perfectly happy. And even if he was not happy, is it bad to feel sad? Is not feeling sad a part of life as well? Do professionals have the right or power to prevent someone from being sad? By not telling him and his family the truth, the possibility to have a normal grieving process and to be able to overcome the pain which comes along with the prognosis is taken from the boy and his family. It got me wondering whether this is a difference in intercultural sensitivity or a lack of knowledge about coping with psychological problems.
>
> The medical staff was very curious about the western way of handling this type of situation. I explained to them that in the Netherlands we try to be very open and honest about a patient's medical situation. The medical staff was very interested in learning more about this – how to tell "bad news", and how to overcome or accept bad news.
>
> After talking to many Vietnamese people about this, it seemed that the approach of the medical staff was based on a combination of lack of knowledge, not knowing how to bring "bad news", and was also culturally linked. They told me that it was perfectly normal for them not to be honest when giving diagnoses or prognoses. Often family members do know the truth, but patients themselves do not. This shocked me, because in my belief, which is similar to the average western belief, every person deserves to know the truth about their medical situation.

This account was chosen for exploration in more depth as it raises the issue of cross-cultural ethics. It also includes a long reflection by the student at the end, when she refers to virtue concepts ("in the Netherlands we try to be very open and honest"). It raises the question for the reader: is honesty a virtue in Vietnam? We have no reason to believe it is not, but in this case, honesty does not appear to have been an issue for the physiotherapists when discussing diagnoses/prognoses with patients until raised by the student. That is, the physiotherapists probably did not see themselves as having a dilemma about whether to be honest or not. However, when this choice was put to them, it appears that respectfulness (towards the family) trumped honesty. Or, as the commentator on this case from Australia suggests, perhaps happiness (of the family) was valued over honesty (towards the family) (Hugman, 2012, pp. 96–97).

The student's response to this issue is interesting. She neither resorts to ethical imperialism (Dutch-style honesty should be adopted in Vietnam) nor does she succumb to ethical relativism (these are the local norms in Vietnam, and we cannot question them). Instead she engages in open, critical dialogue – eager to learn and understand the position in Vietnam, while also keen to open up new ways of thinking for the Vietnamese physiotherapists to consider. She demonstrates what Elshtain (2006, p. 158) calls "a pluralistic relationality of an internationalist sort". She assumes that she and the physiotherapists have enough in common, notably a shared concern with the flourishing of patients/service users, to enable them to recognise their moral disagreements and engage in critical dialogue.

This, according to Moody-Adams (2002), is how cross-cultural learning takes place. Indeed, she argues that "serious cross-cultural moral disagreement is possible only against a background of basic cross-cultural agreement on a substantial number of fundamental moral judgements and beliefs" (p. 7). Admittedly, the "flourishing of patients/service users" as a shared value is relatively thin – it has evaluative content, but little descriptive content relating to what counts as flourishing. However, it is a starting point that does not assume the incommensurability of conceptual schemes, languages and traditions entailed by a radical form of ethical relativism (Bernstein, 2006). It allows the student to engage critically with different ways of thinking and living. As Bernstein (2006, p. 8, italics in original) comments:

> The facile appeal to incommensurability is indicative of an ethical and practical *failure* – the failure to do the hard work required in order to understand and engage different ways of thinking and living.

Moral pluralism and the hard work of ethical engagement

Bernstein (2006), along with many recent moral philosophers (e.g., Kekes, 1993; Lukes, 2008), argues for an "engaged pluralism", which involves recognising different ways of thinking and living, whilst striving at the same time for dialogue and understanding. He describes this as "a demanding task that requires acknowledgement of genuine plurality, but where we nevertheless seek to engage critically the different ways of thinking and living" (Bernstein, 2006, p. 8). As Lukes (2008, p. 140) argues, "acknowledging the facts of moral diversity and value pluralism does not entail abstention from judging others (and their judging us)". The process of engaging critically with others may be challenging, but is sometimes welcomed by oppressed individuals and groups as helping to redefine agendas and break with past ways of understanding character and behaviour, such as redefining violence against women as abuse rather than discipline. This can, therefore, be a process of welcome *appropriation* of new norms by oppressed people themselves, rather than *imposition* of alien Western values by dominant groups. This is one way to respond to the charge of ethical imperialism identified earlier in the chapter in relation to social work. It is not, however, always easy to judge between welcome appropriation of new values and imposition of alien values, and there may often be disagreement and conflict within nations, cultures and communities. This is a key point made in relation to bioethics by Nie and Fitzgerald (2016), who criticise the radical dichotomisation of East and West, local and universal. Instead, they point to the moral plurality within cultures, arguing for a "transcultural common humanity".

Elshtain (2006), in her Augustinian-inspired discussion of the flow between the concrete household and the entire world, draws on the idea of the pilgrim. The pilgrim is a citizen of some country (rooted), traversing the world with a purpose – which is very different from "merely drifting, crossing boundaries and proclaiming one's cosmopolitanism" (p. 159). This echoes Appiah's (2007) call for a rooted cosmopolitanism, which entails being embedded in a particular history or nation, while also being a citizen of the world. While the Dutch student in Vietnam is not a pilgrim, she certainly displays the characteristics of a cosmopolitan traveller with a purpose.

Concluding comments

I now return to our starting point of the global social work statement of ethical principles, having journeyed through some of the concrete particularities of the ethical challenges experienced by social workers in different countries around the world. This statement, particularly the version adopted by IASSW (2018), attempts to address Western ethical imperialism whilst endorsing universal ethical values for social work. Perhaps the main message to the members of the ethics commissions of IFSW and IASSW is to continue the hard work of ethical engagement and to value and learn from the process of dialogue and debate, rather than aiming for a perfect, consensually produced document. Nevertheless, it might be worth considering whether virtue ethics may offer an alternative vision to deontological and consequentialist-inspired ethical frameworks, and may serve better to promote dialogue within a more commonly accepted discourse based on the character of moral agents. While virtue concepts like honesty, courage, and trustworthiness are "thick" and can only be understood in context, there is a common basis of recognition of the discourse of character and virtue across the world, one which lends itself as a starting point for dialogue about ethical matters.

The lesson I have drawn from studying ethics cases from around the world relates to the complexity of the hard work undertaken by professional practitioners to be ethically good social workers. While social workers infrequently give accounts of themselves in terms of specific virtues (e.g., "I want to be an honest person"), they do talk of themselves as puzzling and struggling over how to see situations in ethical terms – what roles to play, how to live up to professional responsibilities, and working on their emotions and reasoning processes. I call this "ethics work" (Banks, 2016a). The capacity to do this work (which could also be described as "moral labour") I have characterised as "professional ethical wisdom" (Banks, 2018), an overarching virtue in the mould of Aristotelian phronesis (Aristotle, 350 BCE/1954). Phronesis involves the capacity to reason about what is right in particular contexts, taking account of nuances and situational particularities. In this sense, it could be regarded as a relatively thick concept, which is also universal in nature as a practice of critical thinking and decision-making (Kristjánsson, 2015). This is exemplified in Case example 2, wherein the student presents herself as working hard at being an ethical practitioner, both in her retrospective accounts of her actual practice in Vietnam and in the creation of her case story.

Implications for teaching and learning

Use of cases like those presented here for discussion with groups of professional practitioners or students in training can be a useful way of exploring transcultural ethics. While cases are often largely accounts of action, raising questions of right and wrong, they can also be interrogated in terms of the relevant virtues of the actors involved. Hence, cases can be used to shed light on virtue concepts and how, as thick concepts, they must be located geographically and culturally for us to make sense of the characterisation of the professionals involved as honest, caring, courageous, or respectful. Exercises using cases are given in Banks and Nøhr (2012), including inviting students to write their own detailed cases, comment on those of others, and select short episodes to act out the roles of key participants. The latter, involving techniques of participatory theatre, can be particularly powerful in encouraging students to develop empathy by inhabiting the shoes of others as actors, judge critically as observers, and combine both as "spectactors" (Boal, 1992).

Acknowledgements

I am grateful to the case authors, and to Routledge for use of summaries of the longer cases published in Banks and Nøhr (2012).

Note

1 These six virtues, described as "universal across cultures and nations", form the basis of the 24 character strengths now used in the VIA Inventory of Strengths (formerly Values in Action); see https://viacharacter.org/character-strengths.

References

Annas, J. (2016). Learning virtue rules: The issue of thick concepts. In J. Annas, D. Narvaez, & N. Snow (Eds.), *Developing the virtues: Integrating perspectives* (pp. 224–234). New York: Oxford University Press.

Appiah, K. A. (2007). *Cosmopolitanism: Ethics in a world of strangers*. London: Penguin Books.

Aristotle (350 BCE/1954). *The nichomachean ethics of Aristotle* (D. Ross, Trans.). London: Oxford University Press.

Baidoun, M., & Lindsey, J. (2012). Commentary 1 [Case 6.4]. In S. Banks & K. Nøhr (Eds.), *Practising social work ethics around the world: Cases and commentaries* (pp. 226–228). Abingdon: Routledge.

Banks, S. (2012). *Ethics and values in social work* (4th ed.). Basingstoke: Palgrave Macmillan.

Banks, S. (2015). Social work ethics. In J. Wright (Ed.), *International encyclopedia of the social & behavioral sciences* (Vol. 22, 2nd ed., pp. 782–788). Oxford: Elsevier.

Banks, S. (2016a). Everyday ethics in professional life: Social work as ethics work. *Ethics and Social Welfare*, *10*(1), 35–52.

Banks, S. (2016b). Professional integrity. In R. Hugman & J. Carter (Eds.), *Rethinking values and ethics in social work*. Basingstoke: Palgrave Macmillan.

Banks, S. (2018). Practising professional ethical wisdom: The role of "ethics work" in the social welfare field. In D. Carr (Ed.), *Cultivating moral character and virtue in professional practices* (pp. 55–69). London: Routledge.

Banks, S., & Gallagher, A. (2009). *Ethics in professional life: Virtues for health and social care*. Basingstoke: Palgrave Macmillan.

Banks, S., & Nøhr, K. (Eds.). (2012). *Practising social work ethics around the world: Cases and commentaries*. Abingdon: Routledge.

Bernstein, R. (2006). Can we justify universal moral norms? In D. Browning (Ed.), *Universalism vs relativism: Making moral judgements in a changing, pluralistic, and threatening world* (pp. 3–17). Lanham, MD: Rowman and Littlefield.

Boal, A. (1992). *Games for actors and non-actors, translated by Adrian Jackson*. London: Routledge.

Bondi, L., Carr, D., Clark, C., & Clegg, C. (Eds.). (2011). *Towards professional wisdom: Practical deliberation in the people professions*. Farnham: Ashgate.

Carr, D. (Ed.). (2018). *Cultivating moral character and virtue in professional practices*. London: Routledge.

Chitereka, C. (2009). Social work practice in a developing continent: The case of Africa. *Advances in Social Work*, *10*(2), 114–156.

Cline, B. (2015). Moral explanations, thick and thin. *Journal of Ethics and Social Philosophy*, *9*(2). https://doi.org/10.26556/jesp.v9i2.89

Dahlsgaard, K., Petersen, C., & Seligman, M. (2005). Shared virtue: The convergence of valued human strengths across culture and history. *Review of General Psychology*, *9*(3), 203–213.

Elshtain, J. (2006). Concrete levels of being and their political implications. In D. Browning (Ed.), *Universalism vs relativism: Making moral judgements in a changing, pluralistic, and threatening world* (pp. 149–161). Lanham, MD: Rowman and Littlefield.

Flanagan, O. (2017). *The geography of morals: Varieties of moral possibility*. New York: Oxford University Press.

Gadamer, H. G. (2004). *Truth and method*. London: Continuum.

Geertz, C. (1973). *The interpretation of cultures*. New York: Basic Books.

Gray, M., Coates, J., & Yellow Bird, M. (Eds.). (2008). *Indigenous social work around the world: Towards culturally relevant education and practice*. Farnham: Ashgate.

Gray, M., Coates, J., Yellow Bird, M., & Hetherington, T. (Eds.). (2013). *Decolonising social work*. Farnham: Ashgate.

Gyekye, K. (2010). African ethics. *Stanford Encyclopedia of Philosophy*. Retrieved December 2010, from http://plato.stanford.edu/entries/african-ethics/

Houston, S. (2003). Establishing virtue in social work: A response to McBeath and Webb. *British Journal of Social Work*, *33*, 819–824.

Hugman, R. (2010). *Understanding international social work: A critical analysis*. Basingstoke: Palgrave Macmillan.

Hugman, R. (2012). Commentary 2 [Case 3.3]. In S. Banks & K. Nøhr (Eds.), *Practising social work ethics around the world: Cases and commentaries* (pp. 96–98). Abingdon: Routledge.

Hugman, R. (2013). *Culture, values and ethics in social work: Embracing diversity*. London: Routledge.

International Association of Schools of Social Work (IASSW) (2018). *Global social work statement of ethical principles (longer IASSW version)*. Retrieved from https://www.iassw-aiets.org/archive/ethics-in-social-work-statement-of-principles/

International Association of Schools of Social Work & International Federation of Social Workers (2004a). *Ethics in social work, statement of principles*. Berne: IASSW and IFSW.

International Association of Schools of Social Work & International Federation of Social Workers (2004b). *Global standards for the education and training of the social work profession*. IASSW and IFSW. Retrieved from www.iassw-aiets.org/global-standards-for-social-work-education-and-training/

International Association of Schools of Social Work & International Federation of Social Workers (2014). *Global definition of social work*. IASSW and IFSW. Retrieved from www.ifsw.org/what-is-social-work/global-definition-of-social-work/

International Federation of Social Workers (2018). *Global social work statement of ethical principles*. Retrieved from www.ifsw.org/global-social-work-statement-of-ethical-principles/

Jones, D. (2012). Commentary 2 [Case 6.4]. In S. Banks & K. Nøhr (Eds.), *Practising social work ethics around the world: Cases and commentaries* (pp. 228–231). Abingdon: Routledge.

Kekes, J. (1993). *The morality of pluralism*. Princeton: Princeton University Press.

Kristjánsson, K. (2015). Phronesis as an ideal in professional medical ethics: Some preliminary positionings and problematics. *Theoretical Medicine and Bioethics*, *35*(5), 299–320.

Lukes, S. (2008). *Moral relativism*. London: Profile Books.
McBeath, G., & Webb, S. (2002). Virtue ethics and social work: Being lucky, realistic, and not doing one's duty. *British Journal of Social Work*, *32*, 1015–1036.
Midgley, J. (1981). *Professional imperialism*. London: Heinemann.
Midgley, J. (2017). *Welfare for a global era: International perspectives on policy and practice*. Los Angeles, CA: Sage Publications.
Moody-Adams, M. (2002). *Fieldwork in familiar places: Morality, culture, & philosophy*. Cambridge: Harvard University Press.
Nie, J.-B., & Fitzgerald, R. (2016). Connecting the East and the West, the local and the universal: The methodological elements of a transcultural approach to bioethics. *Kennedy Institute of Ethics*, *26*(3), 219–247.
Oakley, J., & Cocking, D. (2001). *Virtue ethics and professional roles*. Cambridge: Cambridge University Press.
Osei-Hwedie, K. (1993). The challenge of social work in Africa: Starting the indigenisation process. *Journal of Social Development in Africa*, *8*(1), 19–30.
Pullen-Sansfacon, A. (2010). Virtue ethics for social work: A new pedagogy for practical reasoning. *Social Work Education*, *29*(4), 402–415.
van Oudenhoven, J. P., et al. (2014). Are virtues national, supranational, or universal? *Springerplus*, *3*(223).
Väyrynen, P. (2013). *The lewd, the rude and the nasty: A study of thick concepts in ethics*. New York: Oxford University Press.
Weaver, H. (2016). Ethics and settler societies: Reflections on social work and indigenous peoples. In R. Hugman & J. Carter (Eds.), *Rethinking values and ethics in social work* (pp. 129–145). Basingstoke: Palgrave Macmillan.
Widdershoven, G. (2005). Interpretation and dialogue in hermeneutic ethics. In R. Ashcroft, A. Lucassen, M. Parker, M. Verkerk, & G. Widdershoven (Eds.), *Case analysis in clinical ethics* (pp. 57–75). Cambridge: Cambridge University Press.
Williams, B. (1985). *Ethics and the limits of philosophy*. London: Fontana Paperbacks and William Collins.
Wulfekühler, H., & Rhodes, M. (2018). Meticulous thoughtfulness: Cultivating practical wisdom in social work. *Ethics and Social Welfare*, *12*(4), 330–347.

13 Social work in a pluralistic society

Realising the universal claim of *phronesis*

Dieter Weber and Heidrun Wulfekühler

Universals come from particulars.

(*Nicomachean Ethics* VI. 1143b5)

Introduction

The aim of this chapter is to show what *phronesis* (practical wisdom) can do in relation to the requirements of social work in conflict situations. This question is particularly relevant in view of the fact that social workers practicing in a pluralistic society face an infinite variety of conflict situations. Social work does not only have to do "translation work" by applying ethical principles to an enormous variety of contingent conflict situations. It is precisely these ethical principles, despite their generality, that can only be of local significance because the place of their education is always in a particular culture and its specific life practice. This means that ethical conflict situations in a *pluralistic* society may make it necessary to further develop these ethical principles themselves.

Phronesis seems very apt not only to do this "translation work", but also to further develop these ethical principles. The reason for this lies in the special form of knowledge of phronesis. As practical knowledge, it differs from theoretical knowledge which Aristotle calls episteme. But it also differs from the manufacturing knowledge of a craftsman; what Aristotle calls techne. While episteme as well as techne represent "prior" knowledge, phronesis is initially not knowledge that we can explicate independently of a concrete conflict situation. We follow it, so to speak, "blindly", as we follow "blindly" the rules of how to use linguistic expressions in a language game (*PI* § 219).[1] Just as a language game is a reflection of a practice of life, so is phronesis. As habitualised knowledge, it is always only performed in the concrete individual case.

According to Gadamer, the practical knowledge of phronesis must be, if it wants to be directed at the concrete situation, to "grasp 'circumstances' in their infinite variety" (Gadamer, 2004, p. 19). In this respect, it represents a "concrete generality" (Gadamer, 2004, p. 19). Here, however, the distinction between two relations becomes relevant. First, the relation between the concrete and the general, which abstracts from the individual case. Second, the relation between the local and universal claim to validity. Admittedly, only ethical principles are always sedimented in phronesis, as they are applied in a culturally shaped life practice. In this respect, they represent a local generality, which should not be confused with the "concrete generality" of Gadamer. At the same time, however, precisely those ethical principles are present in phronesis as habitualised so that they cannot be clearly defined. This has to do with the processual character of phronesis, i.e., it can only form and be "alive" if it is applied in concrete, unforeseeable conflict situations. Genesis and performance of phronesis coincide (*NE* VI. 1142a12–16). So if, according to Gadamer, phronesis is capable of "grasping 'circumstances' in their infinite variety" (Gadamer, 2004, p. 19) in relation to concrete situations, then this is precisely where its universal claim to validity lies, despite the only local generality of the ethical principles it habitualises. This means that phronesis can only do the translation work in concrete conflict situations because only there can it develop itself and the ethical principles habitualised in it. Consequently, its universality lies in something procedural. During the translation work, the good (eudaimonia), which is inscribed in phronesis, takes on concrete form. It does this in finding what is appropriate for a concrete, contingent situation, against the background of life as a whole (*NE* 1144a36; *NE* 1140b12). We call this the "momentary urgent". But in order for phronesis to contain the potency to be performed in *all* other contingent situations and to find what is appropriate, phronesis must contain a "lasting important"[2] that preserves its developmental ability and thereby its universal claim.

To argue this, we assume two premises: (1) phronesis performs its translation work in dialogue or by enabling such a dialogue;[3] and (2) phronesis makes such a dialogue possible because it initiates a learning process in which the alterity of the other, by which we mean her/his irreducible individuality, is recognised.[4]

That is why we first turn to Gadamer. He suggests that dialogue as a praxis[5] is the central place where phronesis is performed (Gadamer, 2004, p. 354f; Gadamer, 2001, p. 24),[6] for dialogue is also the genuine place where ethical problems are dealt with. At the same time, phronesis only enables a dialogue (Dottori, 2012, p. 189). According to Gadamer, a central condition for a dialogue is the recognition of the alterity of the other (Gadamer, 2004, p. 355; Lee, 2008; Cesare, 2009, p. 271f.). If the recognition of the alterity of the other is a central enabling condition of dialogue, and if dialogue is the place where phronesis becomes effective, then it is obvious that the extent phronesis itself contains this capacity should be examined. To make this plausible, we want to show that the three aspects of understanding – "empathy", "recognition", and "authenticity" – enable a dialogue and can be regarded as "subexcellences" of phronesis. To the extent that these aspects of understanding are not realised, the basis for recognising the alterity of the other is missing. This will be shown first using a case study from social work. The case study is based on a real case, which is anonymised and tested in teaching over many years.

Second, we want to trace the capacity for empathy, recognition, and authenticity back to implicit relational knowing; a concept postulated by Stern due to his finely analysed observations of the interaction of primary caregiver and child (Stern, 1995). This is a kind of procedural knowledge that is performed in social situations, especially in a dialogue. The central proposal is that this implicit relational knowing can be interpreted as hexis; i.e., as a state of character underlying phronesis, which is decisive for our capacity for empathy, recognition, and authenticity, and thus a crucial prerequisite for our capacity for dialogue and the ability to recognise the other in his/her alterity.

Third, we propose that the developmental path of phronesis and the implicit relational knowing on which it is based as hexis can itself proceed based on the failure of a dialogue and the reflection of this failure, insofar as it enables a new dialogue.

This will be illustrated once again using the case study. In the end, we hope to show how phronesis can fulfil its universal claim precisely in the course of its performance in each concrete contingent situation.

Empathy, recognition, and authenticity as conditions for a dialogue

Here we want to point out the central conditions for understanding that enables a dialogue and the recognition of the alterity of the other, on the basis of the following case study from social work.[7]

> Time and again, the tenants of a residential complex for the elderly complain to its manager, a social worker, about Ms B., a woman of approximately 55 years old. "She drinks, throws beer bottles at us. At night she screams." Some claim she has assaulted them. "But the peak was when, a few days ago, Ms B. was putting faeces all over our doors." The social worker's attempt to get into conversation with Ms B. failed repeatedly. Finally, she persuades Ms B. to talk to her, promising that this will give her the opportunity to pour out her heart. In an interview, Ms B. complains about her neighbours. "They want to poison me and my cat. They secretly mix poison into my food. Plus, toxic gases come out of my light bulbs as soon as I turn on the light," says Ms B. "Don't you want to see a doctor because of your fears?" asks the social worker carefully. "You can't get me back there, you can't get me back there. They want to poison me there too, they all want to poison me. "But it can't go on like this, we've already been through all this," replies the social worker. "They have me tied up, pumped full of Hallifax (she calls the very potent psychotropic Haloperidol "Hallifax"). But not with me.

Not with me. After that, I got a lot worse. I couldn't breathe through my nose. My skin turned red. I can't walk in the sun anymore. And I got fat. As you can see." "But now you're not well either – without any medication. You must also empathise with the other inhabitants." "They're evil, they want to poison me. But I'll show them. I won't get out of here." "Don't you have any family to stand by you or any other person to help you?" "You mean, do I have a supervisor/guardian? I don't need someone like that to steal my money in the end," Ms B. replies.

Dealing adequately with this situation will depend to a large extent on how well the parties to the conflict are able to communicate with each other, and whether they feel understood. In order to reflect on whether the dialogue is successful, the following three communicative aspects can be helpful: "empathy", "recognition", and "authenticity". Before we analyse the case study on the basis of the three communicative aspects, let us briefly explore each of them.

The aspect of *empathy*: With regard to the phenomenon of empathy, we distinguish between a relatively fast, automatic, unreflective efficient process and a relatively slow, reflective, cognitive effortful, i.e., language-bound, process (Frith & Frith, 2008; Apperly & Butterfill, 2009). Empathy, whether pre-reflective or reflective, includes an affective and a cognitive aspect. It includes the ability to put oneself in another's position, while at the same time distinguishing between others and ourselves. On the basis of this understanding of empathy, one can ask: how well do the interaction partners succeed in empathising with each other pre-reflectively, spontaneously, and intuitively/affectively, but also reflectively and cognitively?

The aspect of *recognition*: Following Honneth, one can distinguish three forms of recognition (Honneth, 1995): (1) *love*, in the sense of pre-reflective, spontaneous affective participation, through which the other experiences himself/herself as fundamentally recognised in his/her existence as a human counterpart; (2) *right*, in the sense of the respect of rights and duties deriving from legitimate needs without regard for the person, whereby we experience ourselves as a member of a community of law; and (3) *solidarity*, in the sense of the appreciation of the individual qualities and abilities of a person, who thereby experiences him/herself as a valuable member of a community of values, based on solidarity. In addition, the English "to recognise" or the French "reconnaître" contains a further aspect of recognition in the sense of *identifying recognition* (Ricoeur, 2007), which has a descriptive *and* a normative content. The latter is based on the fact that the identifying recognition related to an individual is confronted with the individual's claim to self-description. In short: every identifying recognition must not make an individual the mere object of a description (with the danger of reification). While love is a pre-reflective, affective form of recognition, *law*, *solidarity*, and *identifying recognition* are based on cognitive-reflective processes. As with empathy, the pre-reflective form of recognition, spontaneous affective participation plays a key role, for without this form of recognition, the willingness to respect the rights of others, to value their individual qualities and abilities, and to recognise one's right to self-description drastically decreases.[8]

The aspect of *authenticity*: The concept of authenticity and its meaning are highly controversial (Kristinsson, 2007). This reflects the struggle for the concept of identity and what it should mean. But without a concept of authenticity, at least in the sense of the experience of selfhood that can no longer be deceived, of a self-given existence in the flesh, we could not make the experience of authorship in decisions and actions. We would then not be prepared to let ourselves be held accountable for our actions. Moreover, the talk of recognition of the alterity of another would make no sense if we could not presuppose the authenticity of the other. Finally, the assumption of authenticity of an interaction partner is fundamental for the success of communication. If we do not appear authentic to our counterpart, if our counterpart instead has the impression that we play a role for him/her, s/he can get the feeling that s/he is not taken seriously, that s/he is not recognised as a serious interlocutor or that s/he is even manipulated. Either way, s/he becomes suspicious and distrustful of his/her dialogue partner. The experience of pre-reflective, spontaneous empathy and recognition also plays a key role here. It is an indicator of our truthfulness, of authenticity. In contrast to the reflective affective and cognitive adoption of perspectives, we can hardly fool the other with regard to our pre-reflective, spontaneous empathy[9] and recognition.

The following case study analysis under the aspects of empathy, recognition, and authenticity can only be very sketchy and selective.

The social worker was able to motivate Ms B. to engage in the conversation with the promise that she could pour out her heart. The social worker conceals her motive for taking the complaints of the other residents seriously. When Ms B. "pours out her heart" to the social worker, the social worker does not really respond to her. To the situation described by Ms B., in which she feels threatened existentially, she answers: "Don't you want to see a doctor because of your fears?" She does this cautiously. But she doesn't respond to Ms B. as one might expect if she were to acknowledge the threat described by Ms B. as "real". Even when Ms B. repeats her fears about a possible inpatient placement in a psychiatric ward, the social worker does not really respond to these fears when she answers: "But things cannot go on like this, we have already done it all". There are real, manifold, easily understandable, clues to the fears of Ms B.: The drug Haloperidol, with all its physiological side effects – not to mention the disturbing psychological effects that one subjectively experiences – can be perceived as a form of "poisoning". And who would be willing to switch places with Ms B. when it comes to admission to a psychiatric facility?

When Ms B. refuses to give in to the pressure of the social worker, the social worker only expresses the interests of the other residents of the old people's home. Once again, Ms B. articulates that she feels threatened by the residents. The other residents really want Ms B. to leave. Also, this time the social worker doesn't directly respond to Ms B., even if she seems to take Ms B.'s worries and fears seriously, as she asks: "Don't you have any family to stand by you or any other person to help you?" At the same time, the social worker presumably confronts a dilemma. If Ms B. had someone who could "really" help her understand her subjective experience of her mental illness she probably wouldn't be in a precarious position.

In all the reactions of the social worker to Ms B.'s remarks it can be seen that she is not really willing (or able) to empathise with Ms B.'s psychological condition, her experience. And to the same extent, she does not really recognise the fears expressed by Ms B. about both her current living situation in an old people's home and psychiatric accommodation. This may be due to a lack of knowledge about mental illness, the experience of the mentally ill and their psychiatric accommodation. But there may also have been fears and prejudices against mentally ill people. After all, she does not prove to be honest and "open" towards Ms B. when she motivates Ms B. to talk about the fact that she may pour out her heart, having in mind first and foremost the interests of the other residents of the home.

Conversely, Ms B. does not seem to respond to the well-intentioned advice, hints, and questions of the social worker and acknowledge that the social worker is also worried about her, even if the social worker is giving priority to the interests of the other residents. Due to her mental illness, Ms B. may struggle to take the perspective of the other residents. For example, her perceptions and fears could be reflective of schizophrenia. Imaging methods (Singh et al., 2015) and psychological tests suggest that individuals with schizophrenia demonstrate reduced or limited empathy in various distinguishable domains (e.g., affective empathy, perspective taking) (cf. the meta-analysis of Bonfils, Lysaker, Minor, & Salyers, 2017). This might suggest that Ms B. finds it difficult to empathise with the other residents. Also, Ms B.'s willingness to acknowledge the interests of her flatmates is not discernible. One may assume that this is due to the sickness-related threat situation but also the non-recognition of her needs by the flatmates. In addition, there is a certain moment of truth in Ms B.'s delusions: they reflect the experience of the deep dislike of the flatmates towards her and that they want to get rid of her. It is unclear to what extent Ms B. can be sincere and authentic in view of her mental illness. In addition, self-reports from schizophrenia patients, accompanied by disturbances of body image and the experience of external control, suggest that they do not experience themselves as themselves in acute phases (Fuchs, 2012).

At the end of the conversation, there is no sign of a solution to the conflict, and it may have made the situation worse. Even though the conflicting interests and perspectives became clear, the understanding between the social worker (who also represented the interests of the neighbours) and Ms B. was very poor. The concepts of empathy, recognition, and authenticity as central elements of understanding can function as important factors for a successful dialogue in an ethical conflict, for they serve the recognition of the alterity, the perspective, needs and beliefs of the other.

Implicit relational knowing as the basis of empathy, recognition, and authenticity as the hexis of phronesis

In the following, we do not only want to attribute the ability to be empathic, to recognise others, and to be authentic to implicit relational knowing;[10] a kind of knowledge which Stern postulated as a result of his analysis of mother-child interaction. We also want to make plausible why such implicit relational knowing as a prerequisite for our self- and social relation suggests itself as an attitude on which phronesis is based. Finally, we want to show that implicit relational knowing has an inherent self-reflectiveness that is necessary for a capacity of phronesis to evolve in social interactions in which it is performed.

Implicit relational knowing is procedural in nature (Boston Change Process Study Group, 2010). It is an incorporated knowledge and is formed in our primary relationships, in interaction with the primary caregiver. The way it develops depends on how *empathetic* the primary caregiver is, how well s/he can spontaneously and immediately *recognise* and satisfy the child's needs and that s/he is *authentic* in doing so; in short, how "sensitive" s/he is in the sense of attachment theory. Implicit relational knowing includes "Representations of Interactions that have been Generalized" (RIG) (Stern, 1985). They represent average expectations of "self-with-other" and are plastic, changing continuously in the course of our lives depending on our interaction experiences (Stern, 1985; 2004). In a positive sense, our self-confidence and basic trust is stored in our implicit relational knowing, from which we draw in social interactions, especially in coping with conflicts, or when we explore new things. It is the model for all further social relationships and structures our self-experience in social interactions.

The implicit relational knowing goes without saying and is communicated spontaneously. This means that social interactions automatically lead to re-enactment. Thereby it is continued, reinterpreted, and rewritten. The implicit relational knowing is self-referential according to its logic of development. It works – in both a positive and negative sense – like a self-fulfilling prophecy. If we are "fixed" on a certain implicit relational knowing due to a lack of recognition, we also define others accordingly. We are closed, we are not ourselves, so we are not authentic. We are not ready and able to empathise with the other's perspective and to recognise him/her in his/her situation. This is expressed in clichés, rigid prejudices that lead to rigid relationship patterns, stereotypical role attributions, projections, transmissions, and counter-transferences; that is, it reproduces and confirms itself, whereby corresponding attitudes harden. If, on the other hand, we have a "positive" implicit relational knowing due to positive recognition experience and social self-efficacy, we are open, we do not define others, and we feel authentic. We do not need to play a role for the other, because we know ourselves as we are, recognised and lovable. We are basically ready to get involved with others, to put ourselves in their position, and to acknowledge them in their needs or in themselves, but also to learn new things about ourselves, to discover completely different sides of ourselves.

But since our implicit relational knowing is constantly reinterpreted, continued, and rewritten in every social interaction (but also in self-reflection – presumably also through meditation), it is not only through self-referential bias that only "compulsively" reproduces itself. Rather, every new encounter causes it to point beyond itself, because only in an encounter is it present. This developmental logic of implicit relational knowing corresponds entirely to the character of phronesis, which is only present in the praxis of a dialogue.

We can state that implicit relational knowing forms the basis of our capacity for empathy, recognition and authenticity as fundamental aspects of understanding that we must presuppose in every dialogue. This is not least because implicit relational knowing is formed precisely through the experience of empathy, recognition, and authenticity, originally – and as a model for all further interactions – in the interaction between the primary reference person and infant. Moreover, implicit relational knowing contains a *certain* self-reference, as is also inherent in phronesis. This self-reference consists in the fact that implicit relational knowing, like phronesis, can only behave critically towards itself and develop further in a dialogical scene in the encounter with another. In the next section, we will explore critical self-reference of phronesis and its underlying hexis, as well as the role of explicit reflection.

The development process of phronesis

The failure of a dialogue does not have to be necessarily synonymous with the non-clarification of the ethical conflict, or even its intensification. It can also become an occasion for the further development of phronesis and its underlying hexis; implicit relational knowing.

In order to illustrate the further development of phronesis and its underlying hexis after the failure of the dialogue in the case study, we want to introduce an important distinction with regard to forms of knowledge: *knowing how* and *knowing that*.[11] While *knowing how* functions as a pre-conscious, pre-reflective, non-propositional procedural intercorporeal knowledge in dialogue, *knowing that* represents a conscious, reflective, propositional, and declarative knowledge. If *knowing how* represents a kind of self-evident knowledge for which no explanation is required, it is characteristic of *knowing that* that we can and want to state causes for facts or reasons for actions. If the *knowing how* of phronesis is present as pre-reflective, procedural knowledge in the second-person perspective, the reflective, declarative knowledge of *knowing that* marks the exit from the dialogue and appears in two complementary perspectives: the third-person and the first-person perspectives. *Knowing that* contains an answer to each of the questions: (1) "what was the case?"; (2) "how and as whom did I perceive the other?"; (3) "how can his communication behaviour be interpreted?"; (4) "how and as whom did I experience myself?"; and (5) "how did I behave towards the other?" These questions are linked to the questions: (6) "who do I want to be?"; and (7) "who do I want to be recognized as?"

In the mode of *knowing how*, empathy, recognition, and authenticity are each effective in pre-reflective forms and can hardly be distinguished from one another. All three can only be reflected upon retrospectively. In the reflective form of knowledge, *knowing that*, empathy, recognition, and authenticity show themselves to be more differentiated and stand out clearly from each other, even if they still remain related to each other. With regard to empathy, we then have not only an affective empathy, but also a cognitive empathy, i.e., the ability to put oneself affectively and cognitively into the perspective of the other. With regard to recognition, we can distinguish different forms of reflective declarative propositional recognition. We can ask as whom we have identified and interpreted our counterpart and his/her behaviour, and to what extent, we have done justice to him/her in his/her self-description and self-interpretation. But we can also consider the extent to which we have recognised him/her in his/her legitimate needs and respected his/her rights and duties. Finally, we can clarify to ourselves to what extent we valued him/her in his/her individual qualities and abilities. With regard to authenticity, we can ask whether we were or could be honest with the other or whether we hid behind a role, and thereby recognise and reveal something of ourselves. These questions, which we have posed to ourselves in the reflective and declarative form of *knowing-that* knowledge, can also be addressed theoretically to our counterpart, if we try to understand his/her communication behaviour.

As long as the dialogue succeeds, i.e., the dialogue partners are able to mutually recognise each other in their alterity, they can make the experience of recognising each other anew through the other. Here the implicit relational knowing – and the abilities of empathy, recognition, and authenticity it makes possible – work like a positive self-fulfilling prophecy; i.e., our trust in ourselves and in the other intensifies. As a result, we are more open, empathetic, and authentic in relation to another and more willing to recognise him/her in his/her alterity, because the dialogue with him/her promises rather a confirmation of one's own person and an intensification of one's own self-relationship. But where the dialogue encounters difficulties and where our capacity for empathy, recognition, and authenticity – our implicit relational knowing – reaches its limits, the dialogue breaks down easily. In view of the failure of the dialogue, we feel compelled to translate the *knowing how* into a *knowing that*; i.e., to translate our non-propositional knowledge into a propositional one. This is because we involuntarily ask why: "why did the other behave like this and react like this?" In view of the lack of understanding, we have already spontaneously given explanations for the behaviour of Ms B. and the social worker in the analysis of the case study. Instead of asking why, we instead may question ourselves. For example, "Why did I react as I did in the conversation? Did I do something wrong?" By trying to give answers to these questions, we are translating our *knowing how* into *knowing that*.[12] So we not only equate phronesis with the *knowing how* as procedural, prereflective intercorporeal knowledge (Dreyfus, 2007; McDowell, 2007; Zahavi, 2013), we also want to show that *knowing that* is embedded in the *knowing how* of phronesis.

At this point, it should be remembered that phronesis and its underlying hexis must remain imperfect (*NE* VI. 1140b22). This imperfection could be interpreted as a deficiency if one were to assume that phronesis is a form of knowledge that already exists in advance, such as the knowledge of techne. If, however, one regards phronesis as a procedural knowledge that is only realised in every contingent, concrete social situation, the attribute of perfection for the characterization of phronesis is completely inappropriate. One thus ignores the processual character of phronesis and must negate its significance in terms of enabling good life practice in the unpredictable contingent life situations. In this respect, the "perfection" of phronesis does not consist in a closed knowledge, but precisely in its unclosedness, for it is only this that makes possible its ability to learn and develop, and thus its ability to find what is appropriate for concrete, contingent conflict situations.

If, as a consequence of the failure of a dialogue, we now involuntarily translate *knowing how* into *knowing that* – so to speak as a first step towards the further development of our practical wisdom or implicit relational knowing – we are confronted with the ambivalent nature of *knowing that*.

On the one hand, this translation can help the social worker to explain the limits of the capacity of empathy, recognition, and authenticity, both on her part and on behalf of Ms B. This explanation can make at least one dialogue partner more empathetic and authentic (in the case study, the social worker) and increase her/his ability to recognise the other in her/his alterity. This may be because we can better explain the interactive behaviour of our counterpart and therefore understand and accept the limits of her/his capacity for empathy, recognition, and authenticity on the basis of theoretical knowledge. The theoretical explanations can relate to the communication behaviour of our counterpart (third-person perspective). Theoretical explanations and self-reflective considerations can also help us to question our own perception, empathy, and behaviour (first-person perspective) – for example, with the help of the theory of recognition, communication theories, psychodynamic theories of transmission, and counter-transmission in unresolved conflicts, etc. We do not then blame our counterparts for something that they cannot readily influence. At the same time, this opens up the possibility of reflecting and seeing through the limits of our own communication behaviour – our own fears, prejudices, projections such as transmissions, and counter-transferences – in relation to the other. The hope is that this translation of our previous *knowing how* into *knowing that* paves the way back into the particular dialogue.

In the concrete case, it is conceivable that the social worker attributes her non-acceptance of Ms B.'s perceptions and experiences to her own unacknowledged insecurity and fears of relating to the mentally ill. For example, she can recognise, through appropriate considerations and theoretical explanations, that these fears and insecurities not only cloud her compassion, but also her view of the extent to which Ms B.'s fears are to be taken seriously and to what extent she is to be recognised as responsible. These considerations may make her realise why she was not honest with Ms. B. when she persuaded Ms. B. for a conversation, with the promise that Ms. B. could say what was on her mind.

On the other hand, the perspective of *knowing that*, formulated in the third-person perspective, can also include the danger of reifying the other. We can make the other alone the object of a theoretical explanation, in which we leave the other no more space to explain her/his behaviour. There are thus narrow limits to the recognition of the alterity of the other. This is accompanied by the possibility of falling into the trap of self-righteousness. We then guard ourselves against being questioned and thereby prevent a change of perspective.

The first alternative opens us to a new dialogue. The second makes further dialogue rather unlikely. One can now ask what decides which of the two paths of *knowing that* opens up. Central here is whether there has been an experience of a new *knowing how* in which our implicit relational knowing has expanded; i.e., our trust in ourselves and in the other has increased. Only against this background do we become capable, in the two reflective attitudes of *knowing that*, of questioning our image of the other, as well as our own behaviour in dialogue with him/her, in order to open ourselves to the alterity of the other. This can formulate through another dialogue reflecting and commenting on the first, for example, in the form of supervision.

Universal claim and local performance and genesis of phronesis: a relationship of mutual conditionality

In view of the claim to the universal validity of phronesis, we can state this: what is appropriate in a contingent situation of conflict we call the "*momentary urgent*". In this specific case, this could possibly mean that Ms B. goes to a day clinic, where she takes medication and receives psychotherapeutic treatment, so that she learns to deal with stress better in her everyday life. At the same time, however, she can keep her apartment in the old people's housing complex, as well as her cat. Within the implicit relational knowing, according to its *potency*, the "*lasting important*" is contained. The "*lasting important*" consists in nothing other than the mutual recognition of the alterity of the other in all her/his indeterminacy/openness. It functions teleologically to find the "*momentary urgent*". With regard to the concrete case, the recognition of the alterity of the other is the permanent stimulus to change perspectives, to put oneself in the other's place, as difficult as it is for the social worker, the residents of the home and Ms B. to do so. On the one hand, this means that one does not stop at explanations that unilaterally blame the other side for the failure of a compromise. For example, the social worker could say that because Ms B. is schizophrenic, and without insight into her illness, one does not have to take Ms B.'s statements of will seriously, as they are an expression of her illness. And on the other hand, it does not mean that the social worker protects herself from questioning herself through self-righteousness. Either lack of knowledge or self-righteousness would stand in the way of understanding. Where the permanently important becomes part of phronesis, as a striving for recognition of the alterity of the other, we can experience that even in very difficult conflict situations we become capable of compromise. The *lasting important* thus functions as a probe to find a compromise in which the best possible solution can be found in each concrete situation, on which the parties to the conflict can agree.

Generally speaking, only in the recognition of the alterity of the other can phronesis *follow* the realisation of the blueprint of eudaimonia in *each* concrete individual case. And the recognition of the alterity of the other is precisely the condition for a dialogue as the excellent place where ethical conflicts are dealt with. On the one hand, the "*lasting important*" enables us to perceive (*aisthesis*) and to deliberate (*bouleusis*); what is appropriate in a concrete contingent situation, the *momentary urgent*. On the other hand, a concrete, contingent conflict situation can – via the failure of dialogue – become an occasion for the further development of phronesis and its underlying hexis, implicit relational knowing. Therefore, this is a critical self-reference of phronesis: it must go through the crisis and overcome a failing dialogue. In other words, phronesis and its underlying hexis retain their capacity for development only through their constant performance in concrete dialogue situations as required by conflict situations. We are thus dealing with a relationship of mutual conditionality.

The development of phronesis is necessarily accompanied by the development of the "owner" of this knowledge. For the habitualised knowledge of phronesis is anchored in its "owner" as implicit relational knowing. Just as phronesis is formed in the course of its performance in concrete conflict situations, so too is the owner of the habitualised knowledge of phronesis.[13] The explication/translation of this knowledge as *knowing that*, which is necessary due to a crisis of understanding, represents a detour of understanding in the dialogue about the explanation. In order to become effective, i.e., practical, this *knowing that* must be translated back into a *knowing how*. According to Aristotle, this translation can be understood as consulting with oneself and a friend. In the professional practice of social work, this takes place in collegial case consultation or supervision.[14]

Notes

1 Wittgenstein, L. (1986). *Philosophical investigations*. Oxford: Basil Blackwell. This is quoted here with PI.

2 The theologian Dietrich Ritschl has introduced the distinction between the momentary urgent and the lasting important in the context of the reflection of the possibilities of ecumenism (Ritschl, 1987, pp. 89–91).

3 According to Kristjánsson, 2014, dialogue is not only a Socratic method in moral education, but also plays a key role for Aristotle. Eikeland has added further evidence to highlight the importance of dialogue for moral education, in order to underline the central role of dialogue in ethics and politics (Eikeland, 2016, p. 46; Eikeland, 2008, p. 399f.).

4 "Alterity of the other"/otherness also includes the separation/strangeness of persons (Cavell, 1976, p. 262). Only in acknowledging our separateness alongside the fact that the other person wants us to share something of ourselves with him or her have we adopted the other person's perspective.

5 In this chapter we assume, for reasons of simplicity, a dialogue between two persons or in one person, probably knowing that the use of the term dialogue is much broader in Plato and Aristotle (Eikeland, 2008, p. 231f.). A dialogue always represents a practice in which the dialogue partners themselves are at stake and are transformed (Eikeland, 2008, p. 234f.).

6 This is logical if, like Gallagher, one considers that phronesis represents primarily an intersubjective – more precisely, intercorporeal – knowledge in the sense of *knowing how* (Gallagher, 2007, pp. 201, 210; Gallagher, 2015, p. 142).

7 This case study is based on a real case from the authors. The real actors and institutions were anonymized.

8 We argue here from an epistemological perspective, as Honneth suggests. According to him, every reflective recognition and acknowledgement is preceded by a pre-reflective, elementary acknowledgement. Honneth, for his part, relies on Stanley Cavell's Wittgenstein interpretation, Martin Heidegger's hermeneutic-existential reference to reality, and John Dewey's qualitative thinking (Honneth, 2012).

9 If one assumes that the unreflected type of empathy takes place relatively quickly, automatically, and therefore involuntarily, and the associated form of pre-reflective recognition as well, it will be difficult to consciously pretend something to someone else. From a neurobiological perspective, it seems that both unreflective and reflective forms of empathy are operating separately, and at the same time are interrelated (Healey & Grossman, 2018; see also Stietz, Jauk, Krach, & Kanske, 2019).

10 In particular, Fuchs has referred our assets of empathy and the assumption of perspectives back to implicit relational knowing (Fuchs, 2016, p. 196).

11 Cf. Gallagher, 2007, 2015 and the Dreyfus-McDowell debate; see Schear, 2013 and Wild, 2014 and further literature there.

12 The reflection on the change of perspective from the participating pre-reflection perspective to a second-order observation is a metacognition. The second-order observation described in the course of metacognition in the sense of third-person and first-person perspectives is a reflective form of social cognition.
13 It should be recalled that our implicit relational knowing can be modified during our whole life (Stern, 2004).
14 In this context, it should be noted that there are initial indications that meditation training can promote the capacity for empathy (cf. Valk et al., 2017).

References

Apperly, I. A., & Butterfill, S. A. (2009). Do humans have two-systems to track beliefs and belief-like states? *Psychological Review*, *116*, 953–970.

Aristotle (2011). Nicomachean ethics. In S. Broadie & C. Rowe (Eds.), Oxford: Oxford University Press.

Bonfils, K. A., Lysaker, P. H., Minor, K. S., & Salyers, M. P. (2017). Empathy in schizophrenia: A meta-analysis of the interpersonal reactivity index. *Psychiatry Research*, *249*, 293–303.

Boston Change Process Study Group (2010). *Change in psychotherapy: A unifying paradigm*. New York: W. W. Norton & Co.

Cavell, S. (1976). *Must we mean what we say? A book of essays*. Cambridge: Cambridge University Press.

Cesare, D. D (2009). *Gadamer – ein philosophisches Porträt*. Tübingen: Mohr Siebeck.

Dottori, R. (2012). Die hermeneutische Erfahrung zwischen Weltoffenheit und Dialog. *Analiza i Egzystencja*, *19*, 177–198.

Dreyfus, H. L. (2007). The return of the myth of the mental. *Inquiry*, *50*(4), 352–365.

Eikeland, O. (2008). *The ways of Aristotle: Aristotelian phrónêsis, Aristotelian philosophy of dialogue, and action research*. Bern: Peter Lang.

Eikeland, O. (2016). If *phrónêsis* does not develop and define virtue as its own deliberative goal – what does? *Labyrinth*, *18*(2), 27–49.

Frith, C. D., & Frith, U. (2008). Implicit and explicit processes in social cognition. *Neuron*, *60*, 503–510.

Fuchs, T. (2012). Selbst und Schizophrenie. *Deutsche Zeitschrift für Philosophie*, *60*(6), 887–901.

Fuchs, T. (2016). Intercorporeality and interaffectivity. *Phenomenology and Mind*, *11*, 194–209.

Gadamer, H.-G. (2001). *Die Lektion des Jahrhunderts: Ein philosophischer Dialog von Ricardo Dottori*. Münster: Lit.

Gadamer, H.-G. (2004). *Truth and method* (2nd ed., J. Weinsheimer and D. G. Marshall, Trans.). New York: Continuum.

Gallagher, S. (2007). Moral agency, self-consciousness, and practical wisdom. *Journal of Consciousness Studies*, *14*(5–6), 199–223.

Gallagher, S. (2015). The practice of thinking: Between Dreyfus and McDowell. In T. Breyer & C. Gutland (Eds.), *Phenomenology of thinking: Philosophical investigations into the character of cognitive experiences* (pp. 134–146). New York: Routledge.

Healey, M. L., & Grossman, M. (2018). Cognitive and affective perspective-taking: Evidence for shared and dissociable anatomical substrates. *Frontiers in Neurology*, *9*, 491.

Honneth, A. (1995). *Struggle for recognition: The moral grammar of social conflicts.* Cambridge: MIT Press.

Honneth, A. (2012). *Reification: A new look at an old idea.* Oxford: Oxford University Press.

Kristinsson, S. (2007). Authenticity, identity, and fidelity to self. In T. Rønnow-Rasmussen, B. Petersson, J. Josefsson, & D. Egonsson (Eds.), *Hommage à Wlodek; 60 philosophical papers dedicated to Wlodek Rabinowicz, department of philosophy, Lund university.* Retrieved from www.fil.lu.sek/hommageawlodek

Kristjánsson, K. (2014). On the old saw that dialogue is a Socratic but not an Aristotelian method of moral education. *Educational Theory*, *64*(4), 333–348.

Lee, K.-B. (2008). *Anerkennung durch Dialog: Zur ethischen Grundlage des Verstehens in Gadamers Hermeneutik.* Kassel: University Press.

McDowell, J. (2007). What myth? *Inquiry*, *50*(4), 338–351.

Ricoeur, P. (2007). *The course of recognition.* Cambridge: Harvard University Press.

Ritschl, D. (1987). *The logic of theology: A brief account of the relationship between basic concepts in theology.* Philadelphia: Fortress Press.

Schear, J. K. (Ed.). (2013). *Mind, reason, and being-in-the-world: The McDowell-Dreyfus debate.* London and New York: Routledge.

Singh, S., Modi, S., Goyal, S., Kaur, P., Singh, N., Bhatia, T., . . . Khushu, S. (2015). Functional and structural abnormalities associated with empathy in patients with schizophrenia: An fMRI and VBM study. *Journal of Biosciences*, *40*(2), 355–364.

Stern, D. N. (1985). *The interpersonal world of the infant: A view from psychoanalysis and developmental psychology.* New York: Basic Books.

Stern, D. N. (1995). *The motherhood constellation: A unified view of parent-infant psychotherapy.* New York: Basic Books.

Stern, D. N. (2004). *The present moment in psychotherapy and everyday life.* New York: W. W. Norton & Co.

Stietz, J., Jauk, E., Krach, S., & Kanske, P. (2019). Dissociating empathy from perspective-taking: Evidence from intra- and inter-individual differences research. *Frontiers Psychiatry*, *10*. https://doi.org/10.3389/fpsyt.2019.00126

Valk, S. L., Bernhardt, B. C., Trautwein, F.-M., Böckler, A., Kanske, P., Guizard, N., . . . Singer, T. (2017). Structural plasticity of the social brain: Differential change after socio-affective and cognitive mental training. *Science Advances*, *3*(10). https://doi.org/10.1126/sciadv.1700489.

Wild, M. (2014). McDowell aus der Sicht der Phänomenologie. In C. Barth & D. Lauer (Eds.), *Die Philosophie John McDowells: Ein Handbuch* (pp. 283–304). Münster: Mentis.

Wittgenstein, L. (1986). *Philosophical Investigations.* Oxford: Basil Blackwell.

Zahavi, D. (2013). Mindedness, mindlessness, and first-person authority. In J. K. Schear (Ed.), *Mind, reason, and being-in-the-world: The McDowell-Dreyfus debate* (pp. 320–343). New York and London: Routledge.

14 Local virtues

The case of the circus

Ron Beadle and Angus Robson

This chapter considers the travelling circus as a context that may help us to consider the relationship between "local" and "universal" virtues. We begin by establishing the relationship between this issue and features of our understanding of the virtues more widely. We draw material from an ongoing contemporary study of circus careers[1] and historic first- and third-person accounts before offering some provisional conclusions.

The title of this volume suggests two possible positions. In the first, virtues are regarded as universal; for some human quality to count as a virtue of any, it must potentially be a virtue of all. In the second, the virtues are "local". According to accounts of this type, it is unproblematic for our shared conception of the virtues that the particular virtues to which you give your allegiance – say a list that includes faith, hope, and charity – and those to which I give mine – a list that includes pride, productiveness, and rationality – comprise different or even exclusive sets. So, what sorts of items should fall within the virtue basket – and how wide or narrow should we consider its brim? To answer this requires us to establish our position on at least four preliminary questions.

First, we must identify the level at which we intend virtues to apply. Whilst most virtue traditions have held the virtues to describe qualities of individual agents, we now encounter a wide variety of claims about the virtues of organisations, teams, and other groupings (see for examples Rego, Nueza, & Cunha, 2010; Neubert, Carlson, Kacmar, Roberts, & Chonko, 2009; Palanski, Kahai, & Yammarino, 2011). However, attribution of virtues at the group level is contentious and risks severing the connection between virtues and character (Cordell, 2017). For the purposes of this chapter, we understand the virtues as naming only the excellences of individual people.

Second, to know what kinds of things might fill the basket, we need a conception of individual characteristics as such. Though vulnerable to the situationist critique (Doris, 2002), the notion that agents might exhibit persistent characteristics in their conduct appears to be presupposed by any account of the virtues. If agents were incapable of acting consistently, then the notion of a characteristic (whether a virtue or anything else) would lose dimensions of temporality and intensity, and, thus vacated, would leave little of interest behind. This chapter presupposes that there are attributes by which agents may be characterised and that the virtues name one such set.

Third, having claimed that the virtues pertain to individuals, the persistence of whose attributes enable their identification, we must now determine how to distinguish between characteristics that comprise virtues and others that do not. Without such a distinction – one in which "humour" may be considered a virtue but "height" not, or in which "courage" may be regarded as a virtue but "extroversion" not – we will be unable to distinguish virtues and vices as a meaningful sub-set of characteristics. The understanding of some or other characteristic as serving and in part constituting the "good" varies with understandings of "good" and its cognates more widely.

Fourth, having established a coherent notion of the types of conditions that must be satisfied to designate a particular characteristic as a virtue, we can only put that concept to use if we can agree an account of the type of evidence that might satisfy these conditions. Here, Alzola (2015) and MacIntyre (1999) have provided contemporary and strikingly similar neo-Aristotelian accounts of the types of intentions, reasons, feelings, and observational conditions under which such attributions might be made. These differ markedly from the kind of conditions that satisfy positive psychology and other non-normative accounts of human character (Beadle, Sison, & Fontrodona, 2015).

As a matter of historical fact, different traditions have given their allegiance to different sets of virtues – heroic virtues, Confucian virtues, Christian virtues, Lakota Sioux virtues, and myriad others have provided distinct social orders with normative orientation. For example, a heroic social order in which virtues were understood as excellences peculiar to social roles occupied by men would have no conception of virtues available to human beings *qua* human beings; the question of local and universal virtues presupposes an understanding of universality that has no place in such a schema (MacIntyre, 1988). It follows that although our contemporaries have developed core conceptions of the virtues, these may not be capable of translation into the idioms and practices of radically divergent social orders. Is a common conception defensible in the face of such diversity? Two different procedures for resolving this issue have been especially influential, those of Peterson and Seligman (2004) and MacIntyre (2007).

Peterson and Seligman's (2004) oft-cited list of character strengths was developed through a conceptually straightforward, though doubtless time-consuming, exercise. Having accumulated items derived from myriad traditions, calculations allowed the development of a set of core character strengths, the Virtues in Action list. Although their source material may have distinguished virtues from vices and non-normative characteristics, the exercise itself involved no normative criteria (Robson, 2015). Were cannibal communities to have shared a conception of a virtue of discerning the succulence of human flesh, then the presence of a sufficient number of such communities in the sample would have led to its inclusion (see also Banicki, 2014).

In place of such accounts, this paper deploys Alasdair MacIntyre's conception of the virtues. In *After Virtue* (2007, Chapter 14), MacIntyre describes radically conflicting sets of virtues at home in the heroic societies, the early Christian church, and nineteenth-century America. He argues that although the characteristics held to be virtues differ widely between these traditions, they nonetheless share an understanding of what it is that distinguishes virtues from both vices and such non-normative characteristics as skills. On MacIntyre's account, virtues are character traits that facilitate the achievement of goods at three interconnected levels – those of practices, of human lives understood in narrative terms and of the moral traditions of communities (MacIntyre, 2007). A characteristic may only count as a virtue if it can be shown to be beneficial in all three domains (MacIntyre, 1994).

Using this understanding of the virtues, a small set of items are claimed as universal inasmuch as no intelligible account of a communal moral tradition, of the good life or the goods internal to particular practices could be envisaged that neither required nor developed the virtues of justice, courage, truthfulness and, in his words, "perhaps some others" (MacIntyre, 2007, p. 192). Further characteristics derive their status as virtues within traditions whose practices, social orders, and self-understandings vary widely.

This chapter uses both historic sources and interview data, collected in Europe and North America during 2018, to exemplify the importance of such virtues in the self-understanding of circus artistes. Its evidence will be merely illustrative and is deployed in support of three theses.

The first is that we must distinguish the different senses in which the "local" and "universal" may be understood if we are to consider the relationship between them. Circus sheds a particular light on this issue because it exemplifies a "mode of existence" (Carmeli, 1991, p. 283) in which geographic locality plays no part. The mobility of the travelling circus, often a condition for its continued existence, requires, as the title of Croft-Cook's (1941) book has it, that "the circus has no home".

Second, evidence from the circus suggests that the importance of particular virtues varies between practices. For example, temperance is a matter of survival for flying trapeze artistes, but is relatively unimportant to clowns. However, both the clown and the flyer will require courage in taking risks to protect the common good of the circus tent when threatened by weather or fire (Beadle, 2018). The protection that the Althoff circus afforded to Jewish artistes during the Holocaust provides a notable example of courage and justice alongside some considerable ingenuity (Otte, 2006, pp. 115–116).

On the neo-Aristotelian account, the attribution of virtue may only be made if certain epistemological conditions are met (Alzola, 2015; MacIntyre, 1999). Those who make claims about others' virtues and vices must have grounds to identify the relevant agent's intentions, have established relevant counterfactual conditions, have witnessed appropriate affect, and have considered these features over a range of situations and times. Without ongoing and proximate relationships with others, we cannot make claims about their virtues. MacIntyre (1998, 2016) holds that the virtues can flourish only in particular types of local political community, those in which our knowledge of and mutual accountabilities to one another are such as to enable such judgments. The third thesis is that circus provides such a context.

We turn first to the issue of locality. As a type of institution, the travelling circus dates back to the early nineteenth century (Bouissac, 2018). Since then, the material conditions of travelling and performing, and the precariousness associated with them, have attended the lives of all who are "del circo" (Hickman, 1995, p. 50). Accounts of circus lives demonstrate that loyalty to this identity as "circus people" often trumps other sources of identity such as nation and even family.

We have considerable evidence that those who undertook what are now known as circus arts before the development of "circus" as an institution, were also nomadic. The rope walkers, jugglers, acrobats, jongleurs, minstrels, and animal trainers that appear regularly in accounts of classical civilisations through to the late Middle Ages and early modernity (Wall, 2012; Hippisley-Coxe, 1951) were essentially nomadic, though many became attached to particular courts and developed other institutions including guilds (Daniels, 2011). They fared well in some social orders, but poorly in most. Legislation requiring rope-walkers to use a safety net was introduced to the Roman amphitheatres in the early second century (Lecky, 1895) but despite Aquinas' defence of jongleurs in the eleventh century (Daniels, 2011), European medieval legislation regarded them as vagabonds liable to summary arrest. Their circus descendants (metaphorically and often literally) continued their vulnerable nomadic life, although they were now members of institutions, albeit temporary ones. As late as the 1820s, witch trials were conducted against circus artistes in America (Wall, 2012, p. 192). Even in the late nineteenth and early twentieth centuries, European circus artistes had to collect references from hosts in so-called "conduct books"; without these, they may have been prevented from crossing borders (Eipper, 1931, p. 85). It is perhaps unsurprising that circus people regard contemporary legislative bans as a continuation of centuries-long persecution (Birkett, 2018).

The suspicion with which circus artistes have been viewed is partly explained by their status as strangers. More positive attitudes towards circus artistes were evident in the 20th century. Reflecting their own internationalism, some enlightenment liberals of that era judged circus as something of a moral exemplar. Consider this extract from 1960:

> politicians might find an object lesson in the spectacle of Germans, Frenchmen, Italians, Scandinavians and English artistes living cheek by jowl, their caravans parked closely together in circus tobers, for months and even years at a time in complete and utter harmony.
>
> (McGregor-Morris, 1960, p. 12)

At a more prosaic level, Eipper's ethnographic account of a Swedish circus in 1931 describes the physical environment that conditions the working experience of circus artistes:

> When the artist opens his door, whether in Italy or the Baltic, he sees the stable alley that leads to the ring. Whoever belongs to a circus knows no boundaries, neither of countries nor of language.
>
> (Eipper, 1931, p. 7)

These somewhat idealistic accounts may fail to acknowledge a degree of national identification manifest, for example, in occasional hostility between American circus artistes who originated from the combatants in World War I (Bradna, 1953). It is nevertheless the case that circus itself operates as a powerful identifier that may be more central to self-understanding than ties of family or nationhood:

> Yvonne described herself as being "del circo", which meant that she too was from a circus family. When her grandmother had first married, she told me, her husband's family had despised the circus and made her give it up. But she missed the life so much that soon she left him and married another man, a Guatemalan, who was from the same world as her, and she had gone on to found her own circus in Costa Rica.
>
> (Hickman, 1995, p. 50)

From another century, and country:

> I asked her once what her first language was, her mother tongue. . . . "I guess French, Italian, German, and English come first." She ticks them off on her fingers. "But then there was Spanish, a little Polish, and, oh yeah, Portuguese. And I'm learning Russian, of course." She forgets to count the Swiss dialects she has mastered over the past few years travelling throughout Switzerland.
>
> (Offen, 2010, p. 479)

Our first thesis is that evidence from different regions and eras demonstrates that the nomadic nature of circus work, artistes' routine exchanges across territories and language, and their identity as "circus", belies any notion of geographic "locality". Circus thereby provides a telling counterexample to any contrast between the merely "local" and the "universal"; the issue remaining is whether the particularities of practices and moral traditions undermine claims of universality to virtues.

Our second thesis is that, while we might be able to identify virtues which are generic to the travelling circus as a way of life, the importance of particular virtues does still vary between practices. We will begin with illustrations of virtues whose importance differs in this way, and then consider those that have been highlighted as critical to maintaining circus communities. In both cases, the pursuit of common goods – those internal to practices and those of the community as a whole – distinguish characteristics that are to be understood as virtues from those that are not.

That particular virtues figure in both the self-understanding and observation of individual circus practices is evident from accounts spanning continents and eras. The virtues of fortitude and constancy have never been far from accounts of acrobatics both ground and aerial, both with and without animals. Wall's account of being knocked out whilst training in the French National Circus School illustrates the relationship between acrobatic skill and fortitude and characterises the traditional understanding of its necessity in the circus community itself:

> The accident rattled me. For a week afterward, I couldn't go near the trampoline without feeling fear. But in the longer term it was almost calming. . . . The old circus families had an expression: "eating sawdust". It referred to a performer's first accident and marked the aspirant's introduction to circus life ("Now that you've eaten sawdust, you might, with a lot of work, become an artist," the famous clown Footit's grandmother told him when he fell off his first horse).
>
> (Wall, 2012, p. 132)

Illustrating the constancy that characterises a French wire walker, Stroud's (1999) ethnography isolates this virtue in both practice and domestic contexts. This kind of observational strategy aligns with the requirements that both Alzola (2015) and MacIntyre (1999) hold for the attribution of virtue:

> He made his own props. He had a precisely engineered practice wire set up at the back of the tent and he would run up and down it every day, his arms held out from his shoulders, balancing the body like wings. He polished the wire every day . . . Along the top of the windscreen of his white van was a sticker reading "Bouglione – Le Professional du Cirque". He did not disappoint this assertion, and it seemed to me that the precision of his life was the precision of his weight on the wire. If his caravan looked dirty he would slip in the air.
>
> (Stroud, 1999, p. 292)

A first-person British account[2] in 2018 exhibits both fortitude and constancy:

> I just got the bug for practising the wire and I used to live on the wire and practice and practice and practice, and eventually got the act I love doing which was the stilts across the wire, doing three jumps in and out of hoops, which is quite dangerous, across the wire.

Inasmuch as fortitude and constancy are required if the internal goods of acrobatic practices are to be achieved, accounts of animal training and juggling emphasise the virtue of patience (Clarke, 1936, p. 46). Writing in 1952, Manning-Sanders observed:

> There is that infinite patience, for instance, that thinks nothing of practising for ten years a trick that may take but a few seconds to perform, and of which, when it is done, the spectators may not grasp the skill. But that does not matter-it is the achievement that counts, for the circus, though on the surface it may appear as glamorous and fantastic as a fairy tale, is, intrinsically, a place of real values.
>
> (Manning-Saunders, 1952, p. 336)

Eipper made similar observations in 1931:

> Blacks, light chestnuts, and iron-greys have been selected among many of their brethren as the very best, and Gautier's mastery has educated their inborn abilities to the last degree of perfection. Day after day I have watched him drilling them, during the winter in Stellingen, and on summer circus tours, and have seen with what never-ending patience a man must try to win the confidence of young, nervous, blooded horses.
>
> (Eipper, 1931, p. 74)

Such examples are not only pertinent to the relationship between particular and universal virtues, but provide the kind of examples that have enabled us to introduce the virtues to our students. Unlike conventional employment contexts, the circus involves levels of risk to life (Beadle, 2018) that parallel many of those apparent in classical discussions of virtue. But not only are such "martial virtues" apparent in this unconventional context.

Whilst acrobats, jugglers, and animal trainers might present the same performance regardless of audience, that sensitivity to context which features in accounts of practical wisdom (Dunne, 2005) inheres in the practices such as clowning (Grock, 1957). Adjusting to the particularities of audiences is critical to excellence here. Consider both third- and first-person accounts of this. Writing in 1931 and describing a Swedish clown troupe, the Bronnetts, Eipper records:

> I have made note of as many as forty two clown-acts of theirs, "entries", as they are known in the language of the lot. I saw and heard perhaps as many as a hundred times in English, German, Dutch, French and Swedish. And the dialogue was invariably somewhat differently shaded, suited to the individual situation and audience. The four brothers reached an understanding with one another while they were working, by means of a glance or a gesture, and took the audience as it wished to be taken, one day playing to the gallery with rough jokes, laying it on thick, and the following day, when the better-class public was in the majority, giving it jests more delicately pointed and gestures less grotesque.
>
> (Eipper, 1931, p. 88)

In 2010, McPherson reports the British clown, Bippo, as making similar claims of his practice; practical wisdom is understood as developing over time and with experience:

> When the audience are coming in, I always go up to people and ask them, "How are you? You all right?" If a man just sort of goes "Yep" and moves along quickly I know he won't be a good volunteer. It's about reading your audience, and I think it takes ages to know how to do it.
>
> (McPherson, 2010, p. 66)

So far in this chapter, we have provided examples of particular virtues that are evident in both third- and first-person accounts of practices across different times and places. Whilst the modalities of particular virtues vary in their contexts such that patience in animal training differs from patience in chess, or practical wisdom in clowning differs from that of police enforcement, they are nonetheless identifiable as the same virtues – and in our teaching practice, such contrasts provide useful illustrations of both the conditions that enable the development of virtues and the different dimensions (e.g., of time, effort, and self-awareness) that attend them in differing contexts. Were we to end these illustrations at the level of practice, however, we would barely have introduced the virtues at all, for in classical and contemporary accounts, the virtues are not only characteristics of lives lived well, but require certain qualities of relationship and commitment between members of particular communities. Once again, circus provides a powerful example, for whilst the internal goods of particular circus practices require different virtues to differing degrees, working on a travelling circus requires commitment to undertaking the common tasks of that community, including the labour associated with building up and pulling down tents and transporting equipment. Consider this 2018 account from a circus rigger in the United States contrasting the collaborative working of the circus community to the competitiveness of school sports:

> All through my entire like high school career and any time I was involved in competitive athletics, it wouldn't . . . it never felt . . . it was always competition within the team. And so it never really felt like it was a people working together towards a goal. And then . . . and [Circus Name] or like tent crew of everywhere was the first . . . was like the biggest thing in my life that was like a collaborative effort. And it was – it was that sense of like working together and your competition or your like you know your antagonist or whatever, is the elements, and is time. And is holy s**t we have to be, you know, eighty miles down the road in like three hours and all this stuff is still up, so let's go to it.

A Danish circus musical director makes the same point in another interview from the same research project:

> To live in a caravan in a small community and the whole mentality of making this show happen every day – if workers go home then we all have to help and being in the rain and having to empty your toilet and being stuck on the side of a motorway – it's a really exciting life and I think so many people join the circus and they fall in love with it because it's so different to a normal life . . . it's something that makes me stay with the circus.

This commitment to the common goods of the community is expressed most dramatically in responses to physical threats such as high winds. The reports that follow are made by clown Bippo in 2010, and the circus owner Betram-Mills, writing in the late 1960s:

> If the wind gets up in the night everyone has to get up to put more stakes in to hold the tent down. It's not like the artiste's think: Aw, it's windy. The workers will get up and put more stakes in. Everybody's up – the girls, everybody. Not through anybody asking – through choice. We're all in this together, and the tent is the most important thing in the circus. If we don't have a tent we wouldn't be able to work. So everyone pitches in.
>
> (McPherson, 2010, p. 62)

> For some reason being a part of a circus means something more than just having a job and getting a living, but none of us knows the reason. The ways in which the spirit of co-operation manifests itself are legion and, although we became accustomed to it, it is in times of stress that it is seen more clearly. When the weather is ghastly and the circus has to be dragged out of one quagmire and built up in another overnight; when it rains or blows incessantly the circus people are in trouble and that means all of them. If floodwater is flowing through stables where their horse and elephants will be in a couple of hours it is the job of grooms and those who own horses to divert the water but you may be sure that every acrobat, juggler and aerialist will be there with a pick or shovel digging trenches.
>
> (Bertram-Mills, 1983, p. 56)

Justice expressed through the involvement of everyone in common effort and courage in the face of threats are pre-requisite to the achievement of the goods of political community (MacIntyre, 1998). The circus illustrates both virtues that are associated with particular practices alongside those that condition the continued existence of the community.

Our final thesis is that the day-to-day life of the community, the risks inherent in its work, and its relative immunity from the lures of external goods, qualifies it as the type of association that fosters moral development (MacIntyre, 1998, 2016). A final selection of voices from within circus exhibits a self-understanding replete with such claims. The first is the mid-twentieth–century clown Reynolds:

> There seems to be something stultifying about mere size, and human beings seem to be at their best in the smaller units: the Greek City State, the English country village, the little family business, the small tenting circus were all good forcing grounds for the interesting and curious flowers of human personality.
>
> (Reynolds, 1954, p. 197)

From the same period, the writer turned clown Mardon observed of one of his colleagues, an exceptional acrobat, that:

> He works hard, possibly risking his neck, for little money and no fame; no adulation will come his way, no teenagers will swoon, no mounted police will have to be called out to clear his path; his name needs no neon lights; not for him the brief notoriety of the glossy weeklies, nor the celluloid splendours of a film star's life; he will never be asked to take part in a TV panel game. Not for him the fetishes and superstitions of addle democracy. Like a monk, he had little, yet he has much.
>
> (Mardon, 1961, p. 125)

Both the intensity of workflow and the necessity of teamwork are regarded as requiring the development of virtues in contemporary accounts:

> Aloysia Gavre, the director of Cirque School in Los Angeles, described the phenomenon as "quick bonding," an accelerated version of the teambuilding exercises that most companies foist on their employees. "in almost every circus act," she told me, "you are being asked to trust people quickly, whether you're being thrown, pitched, or caught, whether somebody is holding your line or not offstage."
>
> (Wall, 2012, p. 178)

The mode of institutionalisation of circus practices threatens the somewhat cosy picture that we have drawn in this chapter. In circus as elsewhere, institutional forms and the pursuit of external goods can distract and corrupt but many of the leaders of circus communities understand this threat and their deliberate desire is to maintain their communities on a scale and in a pattern that encourages the kind of communal life that fosters their own and other virtues (Beadle, 2013). It is considerations such as these that have led MacIntyre to endorse the idea that circus is both practice and institution

(MacIntyre 2008, p. 4). The development of social circus and circus school movement makes explicit such communal and political ends. Reg Bolton, who pioneered social circus, wrote this:

> Perhaps the most significant thing about New Circus is where it happens. No longer is circus confined to the Big top, although it is, of course, still happening there too. New Circus can be found in the street, in the community hall, in the park, in the theatre, in cabaret, on TV, in schools and holiday camps.
>
> (Bolton, 1987, p. 7)

Jessica Hentoff, director of Circus Harmony, a circus school in St. Louis, says:

> Circus Harmony teaches the art of life through circus education. We work to build character and expand community for youth of all ages, cultures, abilities and backgrounds. Through teaching and performance of circus arts, we help people defy gravity; soar with confidence, and leap over social barriers, all at the same time.
>
> (Cited in Levinson, 2015, p. 22)

We have argued for three theses on the evidence of circus life presented here. First, the idea of locality in virtues, if interpreted in geographical terms, is misleading and is better understood as particularity. Second, we can understand virtues as being to varying degrees particular to certain practices, understood within the broader context of a moral community. Third, travelling circus is the kind of moral community which is in fact able to foster the development of the virtues.

It does this through the requirements of the virtues for the achievement of the goods of circus practices rather than through formal education; the virtues are learned but not taught. In the traditional circus, children begin to learn circus skills as a form of play (Lowndes, 1962, p. 123), and as they develop their acts, the importance of cooperation and trust (Cottle, 2006, p. 242; Eipper, 1931, pp. 175–176), of regular practice and self-discipline (Mardon, 1961, p. 125; Schreiber, 2005, pp. 132–133), of patience (Croft-Cook, 1941, p. 93; Manning-Saunders, 1952, p. 336), and of justice in the allocation of tasks, honours, and rewards (Bradna, 1953, p. 73; Otte, 2006, p. 55) becomes apparent. Wall (2012, p. 178), reporting on the formal learning in contemporary circus schools emphasises resilience and trust.

What overall conclusions might we reach from these arguments? It appears that the idea of "universal or local" virtues is a misleading dichotomy, not only because the idea of locality is in itself misleading, but also because the dichotomy itself is false. The virtues are, we might say, to some degree universal and to some degree particular. The basic idea that the virtues enable the pursuit of common goods implies some degree of universality, whilst the idea that people develop and exercise virtues in specific practices implies some degree of particularity. The circus provides empirical support for the thesis that virtues are neither particular nor universal but always both. Further, we have proposed that specific communities may (or may not) enable and sustain the virtues of their members, or at least that this kind of sustenance is best understood in the context of specific communities.

This, then, leads to the thought that the term "universal" itself may also be misleading, since it appears to operate at some level which is beyond any community at all. It appears to connect the individual with the absolute, omitting the social structures which we have been exploring as necessary for moral development. We may therefore prefer some other term than "universal", such as "common" or "general". It will then be unproblematic to say that virtues are always both general and particular. Thus, instead of the question, "Are virtues universal or local?" we now have different and perhaps more fruitful questions. If virtues are always to some degree both general and particular, how are people enabled to develop them in the context of specific practices and moral communities? Which kinds of community develop and sustain the virtues, and which do not?

Examples such as the circus, in which the development of virtues is so clearly necessary, provide resources for considering the relationship between the development of virtues and the characteristics of communities. It also provides material for case study and role-based classroom activities that enable students to work beyond familiar settings as they consider a contemporary context which – in terms of scale, risk, and relationships – has more in common with the types of communities in which serious consideration of the role of virtues began.

Notes

1 Meaningful Work and Career Priorities for Creatives in Circus, British Academy – Leverhulme Small Research Grant Award SG162738 (2017–2019).
2 The anonymous 2018 first-person accounts quoted here and following are drawn from the ongoing research project mentioned at the start of this chapter.

References

Alzola, M. (2015). Virtuous persons and virtuous actions. *Business Ethics Quarterly*, *25*(3), 287–318.

Banicki, K. (2014). Positive psychology on character strengths and virtues: A disquieting suggestion. *New Ideas in Psychology*, *33*(1), 21–34.

Beadle, R. (2013). Managerial work in a practice-embodying institution – The role of calling, the virtue of constancy. *Journal of Business Ethics*, *113*(4), 679–690.

Beadle, R. (2018). On running away to the circus. *Politics and Poetics*, *IV*, 1–30.

Beadle, R., Sison, A., & Fontrodona, J. (2015). Introduction: Virtue and virtuousness: When will the twain ever meet? *Business Ethics: A European Review*, *24*(S2), 67–77.

Bertram-Mills, C. (1983). *Bertram mills circus* (2nd ed.). Avon: Ashgrove Press.
Birkett, D. (2018, October 27). In defence of circus animals. *The Spectator*. Retrieved from www.spectator.co.uk/2018/10/in-defence-of-circus-animals/
Bolton, R. (1987). *New circus*. London: Calouste Gulbenkian Foundation.
Bouissac, P. (2018). *The meaning of the circus: The communicative experience of cult, art and awe*. London: Bloomsbury.
Bradna, F. (1953). *The big top – My forty years with the greatest show on earth*. London: Hamish Hamilton.
Carmeli, Y. (1991). Performance and family in the world of British Circus. *Semiotica, 85*(3/4), 257–289.
Clarke, J. S. (1936). *The circus parade*. London: BT Batsford.
Cordell, S. (2017). Group virtues: No great leap forward with collectivism. *Res Publica, 23*(1), 43–59.
Cottle, G. (2006). *Confessions of a showman: My life in the circus*. London: Vision.
Croft-Cooke, R. (1941). *The circus has no home*. Toronto: The Falcon Press.
Daniels, N. (2011). *From jongleur to minstrel: The professionalization of secular musicians in thirteenth- and fourteenth-century Paris*. Charles Singleton Center for the Study of Premodern Europe, Johns Hopkins University. Retrieved from https://jscholarship.library.jhu.edu/handle/1774.2/35244
Doris, J. M. (2002). *Lack of character: Personality and moral behavior*. Cambridge: Cambridge University Press.
Dunne, J. (2005). An intricate fabric: Understanding the rationality of practice. *Pedagogy, Culture and Society, 13*(3), 367–389.
Eipper, P. (1931). *Circus: Men, beasts and the joys of the road* (F. H. Martin, Trans.). London: George Routledge and Sons.
Grock (1957). *King of clowns* (B. Creighton, Trans.). London: Methuen.
Hickman, K. (1995). *A trip to the light fantastic: Travels with a Mexican circus*. London: Flamingo.
Hippisley-Coxe, A. (1951). *A seat at the circus*. London: Evans Brothers.
Lecky, W. E. H. (1895). *A history of European morals*. New York: Appleton and Co.
Levinson, C. (2015). *Watch out for flying kids: How two circuses, two countries, and nine kids confront conflict and build community*. Atlanta, GA: Peachtree.
Lowndes, J. S. (1962). *How the circus works*. London: Routledge and Kegan Paul.
MacIntyre, A. (1988). *Whose justice? Which rationality?* London: Duckworth.
MacIntyre, A. (1994). A partial response to my critics. In J. Horton & S. Mendus (Eds.), *After MacIntyre: Critical perspectives on the work of Alasdair Maintyre* (pp. 283–304). Notre Dame: University of Notre Dame Press.
MacIntyre, A. (1998). Politics, philosophy and the common good. In K. Knight (Ed.), *The MacIntyre reader* (pp. 235–254). Cambridge: Polity Press.
MacIntyre, A. (1999). The recovery of moral agency. *Harvard Divinity Bulletin*, April 16, 1999; Reprinted in Wilson, J. (ed.). (2000). *The best Christian writing* (pp. 111–136). London: HarperCollins.
MacIntyre, A. (2007). *After virtue: A study in moral theory* (3rd ed.). Notre Dame: University of Notre Dame Press.
MacIntyre, A. (2008). How Aristotelianism can become revolutionary: Ethics, Utopia and resistance. *Philosophy of Management*, 7(1), 3–8.
MacIntyre, A. (2016). *Ethics in the conflicts of modernity – An essay on desire, practical reasoning and narrative*. Cambridge: Cambridge University Press.
Manning-Saunders, R. (1952). *The English circus*. London: Werner Laurie.

Mardon, M. (1961). *A circus year*. London: Putnam.
McGregor-Morris, P. (1960). *Sawdust and spangles*. London: H. F. & G. Witherby.
McPherson, D. (2010). *Circus mania*. London: Peter Owen.
Neubert, M. J., Carlson, D. S., Kacmar, K. M., Roberts, J. A., & Chonko, L. B. (2009). The virtuous influence of ethical leadership behaviour: Evidence from the field. *Journal of Business Ethics*, *90*(2), 157–170.
Offen, J. L. (2010). Portrait of a circus girl. *Ethnography*, *11*(3), 473–484.
Otte, M. (2006). *Jewish identities in German popular entertainment 1890–1933*. Cambridge: Cambridge University Press.
Palanski, M., Kahai, S., & Yammarino, F. (2011). Team virtues and performance: An examination of transparency, behavioral integrity, and trust. *Journal of Business Ethics*, *99*(2), 201–221.
Peterson, C., & Seligman, M. E. P. (2004). *Character strengths and virtues: A handbook and classification*. Oxford and Washington, DC: Oxford University Press; American Psychological Association.
Rego, A., Nueza, R., & Cunha, M. P. (2010). Perceptions of organizational virtuousness and happiness and predictors of organizational citizenship behaviors. *Journal of Business Ethics*, *93*(2), 215–235.
Reynolds, B. (1954). *Broken hearted clown*. London: Arco Publications.
Robson, A. (2015). Constancy and integrity: (Un)measurable virtues? *Business Ethics: A European Review*, *24*(S2), 115–129.
Schreiber, M. (2005). *Dreams of the solo trapeze: Offstage with cirque du soleil*. Austin, TX: Greenleaf Book Group.
Stroud, N. (1999). *Josser: Days and nights at the circus*. London: Little Brown and Co.
Wall, D. (2012). *The ordinary acrobat: A journey into the wondrous world of circus, past and present*. New York: Alfred A Knopf.

15 Confucian virtue ethics and Confucian character education

Yi-Lin Chen

Introduction

The aim of this chapter is to elaborate on the idea that the primary objective of Confucius' ethics is to make people virtuous, and to this end, ritual education imbued with the education of emotions is the linchpin of Confucian character education. Premised on the idea that both Confucius and Aristotle share a virtue approach to ethics (Yu, 2007), some similarities between their respective character education would also be indicated.

The Analects (pronounced *Lunyu*, hereafter *AN* for short) means literally the ordered sayings of Master Kong. It was generally acknowledged to have been put together after the death of Confucius (551–479 BC). It comprises conversations between the Master and his disciples, anecdotes about the Master's conduct in various situations, and selected admonitions from the mouths of the Master and some of his disciples (Slingerland, 2001). *AN* is a critical text for anyone who intends to understand Confucius' ethics. Although it is not a formal treatise on ethics, it is by no means simply aphoristic; rather, it contains an implicit conceptual framework for ethics (Mahood, 1974; Olberding, 2012).

This volume consists of 20 books, which comprise 499 chapters. The major topics include what is *ren*仁(as a general virtue, perfect virtue), how to be a *junzi*君子/gentleman (namely, a *ren* person仁人), rich descriptions of *li*禮/ritual propriety and xiao孝/filial piety, and the significance of making friends友 and role models for cultivating virtue. On the whole, *AN* is centred chiefly on moral self-cultivation修己. And the ideal personality or a fully realised human being in Confucianism is represented by the notion of "gentleman" (*junzi*) (Wong, 2011).

Virtuous person and virtuous official

Luo makes a strong case that *AN* has abundant textual evidence showing that "the primary objective of Confucius' teachings is to make his disciples fit to take part in government" (2012, p. 17). When Tszu-lu asked about the gentleman, Confucius said, "He cultivates himself in reverence", "He cultivates himself so as to give rest to others [his clan]", and "He cultivates himself so as to give rest to all the people [in the empire] – even Yao and Shun found it difficult" (Legge, 2015 [*AN* 14.43]). For that matter, *AN* is not only concerned with how to become a virtuous *person*, but also how to become a virtuous *leader* serving the government and practising virtue politics (Luo, 2012). This is because in that disturbing time, a *ren* government official could have much greater influence on society than merely a virtuous commoner (Luo, 2012). This is in line with Confucius' attitude towards social unrest. He does not want to be a virtuous recluse; instead, he said "Sell it! Sell it! I am one waiting for such a merchant" (*AN* 9.13). He said to Yen Yuan, "When employed, to put it [the Way] into practice; when unemployed, to keep it in store – perhaps only you and I are capable of this" (*AN* 7.11).

The virtuous official is expected to act as a role model for his subordinates, and in so doing, he takes the responsibility of educating his people to be virtuous, as well. This is in tune with Confucius' idea of *ren*/benevolence/humanity to the effect that "It is to love all men" (*AN* 12.22). "If the gentleman acquires the Way, he loves men" (*AN* 17.3). "Can you [an official] love them without making them toil? Can you be loyal to him [the sovereign] without admonishing him?" (*AN* 14.7). The *zen* person is one who, wishing to establish himself, seeks also to establish others; he who wishes to gain perception, i.e., to gain a thorough understanding of the Way of humanity (Huang, 1997), helps others to gain perception (*AN* 6.30). For that matter, *AN* is a treatise not only on ethics, but also on politics and education. This volume as a whole is, indeed, mainly concerned with character education, both of a would-be *ren* person's own moral self-cultivation and that of his subjects.

Confucius put supreme emphasis on the virtuous official's role of acting as a model for his people. When Ji Kang-zi asked Master Kong about government, he replied: "Government means correctness/rectification. If you take the lead in being correct, sir, who dares remain incorrect?" (*AN* 12.17). Master Kong also said, "If you yourself are correct, even without the issuing of orders, things will get done; if you yourself are incorrect, although orders are issued, they will not be obeyed" (*AN* 13.6). "If you can set yourself correct, what difficulty do you have in conducting state affairs? If you cannot set yourself correct, how can you correct others?" (*AN* 13.13). Clearly, for Confucius, state affairs have to do with correcting people, and such correction is the official's job.

To this end, the official must first make himself moral and become virtuous/*ren*. That is because people are always watching the example set by their rulers, who can act as lively models for imitation. For that matter, Confucius repeatedly admonished his disciples who were expectedly would-be virtuous leaders that: "If you desire goodness, the people will be good accordingly. The gentleman's moral character is wind and the small man's [i.e., the common people's] moral character, grass. When the grass is visited by the wind, it must surely bend" (*AN* 12.19). The people's character is likely to be corrected by the virtuous leader's moral character. Besides, the other side to the same coin is this: Zi-gong said: "The gentleman's errors are like the eclipses of the sun and the moon. When he makes one, everyone sees it; when he corrects it, everyone looks up to him" (*AN* 19.21). In sum, virtuous leaders are expected to be moral exemplars for the common people to look for, to imitate, and to emulate. The educational significance both of modelling and emulation is highly appreciated by Confucius (see also *AN* 4.17). This is in line with Aristotle's stress on the educational value of the emotional virtue of "emulation", especially for the young (Aristotle, 2004; Kristjánsson, 2006, 2007). Aristotle's sophisticated conceptual analysis of emulation helps to complement Confucius' idea. Since *AN* is concerned with becoming virtuous, a follow-up question is how to do this? Regarding this, a series of core concepts in *AN* come to the fore including *ren*, *li*, and *yi*義.

Ren as the ideal character is applicable to all

Wing-Tsit Chan, in his seminal paper entitled "The evolution of the Confucian concept *Jên*", indicates that "*jên* is essentially a Confucian concept, and it was Confucius (551–479 [BC]) who made it really significant" (1955, p. 295). In the pre-Confucian texts, *jên/ren* is mentioned only occasionally, and in all cases it denotes "the particular virtue of kindness, more especially the kindness of a ruler to his subjects" (Chan, 1955, p. 295). In stark contrast to that pre-Confucian understanding, *jên* in Confucius' usage undergoes a radical change. Specifically, three points are remarkable (Chan, 1955): first of all, *jên* is made a central theme of *AN*. Fifty-eight of the 499 chapters (11.6%) are devoted to the discussion of *jên*, and it appears 105 times. Second, Confucius transformed the ancient meaning of *jên* as a particular virtue into what was appropriately translated as "perfect virtue"; that is, *jên* in the sense of being an inclusive virtue, moral life at its best. Finally, *jên* is no longer a special characteristic of rulers but "a quality applicable to all human beings" (Chan, 1955, p. 299). In the context of the current volume this means, as we see following, that Confucius firmly believed in the universality of virtue.

For Confucius, *ren* is not the monopoly of highborn people. *Ren* can be attained by all people. The ideal character of "a complete man" 成人 (*AN* 14.12) is assumed to be valid for all people. That is, insofar as the ideal character is concerned, for Confucius, there is no distinction between individuals based on their status as in Aristotle. This is termed "the claim to universality"; "an appeal to universality is part of the structural vision of *the Analects*" (Slingerland, 2001, p. 117). At the outset, it must be noted that *ren* has been translated into many different English terms, and this fact shows that *ren* is an exceedingly complicated concept (Chan, 1955). In general, *ren* in Confucius' usage has two meanings. One is *ren* as a particular virtue. In this narrow sense, *ren* is best translated as "benevolence" and it is often contrasted with wisdom and courage (e.g., *AN* 9.29, 14.29) (Chan, 1955; Cua, 1971) and mentioned along with being clever, trustworthy, forthright, courageous, unbending, and so on (Yu, 1998, 2007). The other is *ren* as perfect virtue and it acts as "an articulation and specification in the context of a moral ideal of Confucius" (Cua, 1971, p. 126). *Ren* in the latter sense includes a set of particular virtues such as filial piety, wisdom, truthfulness, courage, loyalty, earnestness, liberality, truthfulness, diligence, and generosity (*AN* 17.5).

Ren, for Confucius, is a distinguishing potential of human beings. As a potential common to all, it awaits to be realised. As put, "Is *ren*/humanity so remote? If I desire *ren*/humanity, there comes *ren*/humanity!" (*AN* 7.30). The root of *ren* is inside everyone, be he of noble birth or from a lower social stratum (bear in mind that many of Confucius' disciples were from the lower classes). In reply to Yen Yuan's question about perfect virtue, Confucius asks a rhetorical question, "The practice of *ren*/humanity rests with oneself. Does it rest with anyone else?" (*AN* 12.1). *Ren* is not inborn, but awaits to be cultivated and realised to the full. It is not beyond anyone's reach. "Is there anyone who can, for a single day, exert his energy on *ren*/humanity? I have never seen any whose ability is insufficient. There may be such people; only I have not seen any" (*AN* 4.6). As a consequence, when Zan Ch'iu said, "Not that I do not like your Way, sir, but that my ability is insufficient", the Master admonished him that: "Those whose ability is insufficient give up halfway. Now you have drawn a halting line" (*AN* 6.12).

That said, Confucius did not believe that human nature is naturally attracted to *ren* (Yu, 1998). A straightforward example is this: the Master said, "I have not seen a person who [naturally] loved virtue, or one who [naturally] hated what was not virtuous" (*AN* 4.6). Also, the Master said, "I have not seen one who loves virtue as he loves beauty" (*AN* 9.18). For that matter, Confucius claimed that *ren* is "to overcome oneself and to return to *li*" (*AN* 12.1). The enterprise of overcoming oneself has to do with disciplining one's natural inclinations. This seems to echo Aristotle's endorsement of Plato's idea that "We ought to have been brought up in a particular way from our very youth . . . so as both to delight in and to be pained by the things that we ought; this is the right education" (Aristotle, 2009, p. 26). That is, the cultivation of *ren* is mainly concerned with educating the emotions and natural inclinations (discussed later).

As noted previously, *ren* is not the private property of aristocrats; it is open to all, irrespective of birth, social rank, or intelligence. That is because "By nature, people are close to one another; through practice, they drift far apart" (*AN* 17.2). Furthermore, some textual evidences indicate that *ren* as an ideal character is considered to be suitable for all peoples. For instance, when Fan Chi asked about *ren*/humanity, the Master said: "Conduct yourself with respect; perform your duties with reverence; treat others with wholehearted sincerity. Even if you should journey to the Yis and Dis, you cannot abandon these" (*AN* 13.19). Yi and Di are ancient names for the non-Chinese tribes, respectively, to the east and the north of China proper. "The Yis and Dis" is utilised to represent all of the non-Chinese tribes on the borders of China (Huang, 1997). *Ren*/humanity will work in foreign countries, too. Confucius once commended these foreign tribes for respecting their rulers, saying "Even the Yis and Dis have regard for their sovereigns, unlike the various states of Xia, which have none" (*AN* 3.5). In sum, *ren* is an ideal not restricted to the Chinese people.

Bending one's mind on *ren* as a precondition for self-cultivation

For anyone to make effort on self-cultivation, a prerequisite is to set one's will on virtue first (*AN* 4.4). That is the motivational or volitional aspect of self-cultivation. Confucius put great emphasis on determination and setting one's will on learning, virtue, truth, and the Way (e.g., *AN* 1.11, 2.4, 4.4, 4.9, 7.6). That is because: "The three armies may be robbed of their supreme commander, but the will of even a common man cannot be taken from him" (*AN* 9.26). Besides, "The determined scholar and the man of virtue will not seek to live at the expense of injuring their virtue. They will even sacrifice their lives to preserve their virtue complete" (*AN* 15.9).

The foregoing discussion indicates that for Confucius, any person who fixes his will on the achievement of virtue can possibly be a person of *zen*. Confucius' emphasis on determination explains why the notion of "self" cultivation is characteristic of his ethics, but it does not mean that the cultivation of virtue cannot begin until one sets his will on it. For instance, Confucius said, "At fifteen, I had my mind bent on learning" (*AN* 2.4). With that said, it is plausible to think that before Confucius was 15, his caretakers had already begun the enterprise of inculcating virtues in Confucius. It was at 15 that Confucius consciously assumed responsibility for self-cultivation. Confucius' stress on self-education is understandable if we remember that his disciples had set their minds on learning. For that matter, Confucius in *AN* was mainly not in the business of teaching children (Slingerland, 2001).

The primacy of *li*

Regarding Confucius' character education, the best place to start is with *li*/ritual propriety. *Li* is Confucius' greatest preoccupation and the foundation of his virtue curriculum (Sarkissian, 2010). As Flanagan writes, "The practice of *li* is the lynchpin of Confucian ethics" (2011, p. 169). Confucius said to his son, Bo-yu, that "If you do not learn *The Rituals*, you have no way of establishing yourself" (*AN* 16.13). (The idea appears also in *AN* 8.8 and 20.3.) *Li* is the first step to take on learning to be a gentleman.

That is because *li* as a traditional repository of rules guiding personal behaviour and interpersonal relations (Chong, 1998) pervades all areas of life. The rituals act as a code of propriety governing all phases of human life, including human conduct, etiquettes governing family life and social relations, state affairs, rulership, and religious sacrifices (Huang, 1997). The coverage of *li* includes all important affairs involved from birth to death (e.g., *AN* 2.5, and Book 10). No one can lead a decent life or behave properly without following the guidance of *li*, even the sovereign (e.g., *AN* 3.19). The rites work as definite rules regulating people's behaviour in different contexts, be it alone or in relation to others. The Master said to Yan Yuan, "That which does not conform to the rituals – do not look at it; which does not conform to the rituals – do not listen to it; that which does not conform to the rituals – do not say it; that which does not conform to the rituals – do not do it" (*AN* 12.1). Confucius recommended virtue education should begin with the learning of the rites. *Li* is learned mainly through guidance, observation, imitation, and habituation.

Among others, Books 3 and 10 are most directly centred on *li*. Book 10 is usually read as a record of the various ritual norms followed by Confucius and many readers have been amused, intrigued, or baffled by it (Tan, 2005). Two examples are as follows: "He [Confucius] would not sleep like a corpse, nor attend to his appearance at home" (*AN* 10.17); "If the mat was not set right, he would not sit" (*AN* 10.7). In order to dispel the apparent ridiculousness, we need to investigate what Confucius expected exactly of *li* in terms of its function and real value in virtue education.

As a starting point of virtue education, the function of *li* for Confucius is like that of laws for Protagoras. In Plato's *Protagoras*, Protagoras remarked that the state compelled the young to learn the laws with the aim of making them "live after the pattern which they furnish, and not after their own fancies" (Plato, n.d.). Likewise, the rites set up the principles of right acts which are later on utilised to guide the young in their future conduct. In short, *li* is intended to offer explicit codes of behaviour to be followed. Those transgressing *li* are to be corrected.

Confucius took great pains to highlight the "restraining" and "regulating" effect *li* has on human behaviour (as well as human psychology, as I suggest later). "A gentleman . . . [who] restrains himself with the rituals is not likely to betray" (*AN* 6.27, 12.15). His distinguished disciple Yan Yuan once praised Confucius for being good at "restraining me with the rituals" (*AN* 9.11). Also, in replying to Yan Yuan asking about *ren*/humanity, the Master said, "To restrain oneself and return to the rituals constitutes *ren*/humanity" (*AN* 12.1). Underlying Confucius' practice is his belief that: "Those who err through self-constraint are rare indeed" (*AN* 4.23). That is because *li* represents the guiding rules preventing people from error and directing people towards right behaviour. Apparently, the restraining effect of *li* is on outward performance.

Li exerts a restraining effect on human behaviour both at the level of individuals and at the level of state governments. The Master admonished his disciples that "If you govern them [common people] with virtue and regulate them with the rituals, they will have a sense of shame and flock to you". (*AN* 2.3) The rituals are expected to work as a worthwhile educational measure that virtuous leaders always take. The worthwhileness is understood both in terms of the contents and the restraining effects of the rituals.

The essence of *li*: the proper spirit; the genuine and sincere feelings

G. H. Mahood claims that: "Too many western-oriented philosophers conceive *li* in narrow ritualistic terms, forgetting that Confucius asked: 'Are gems and silk all that is meant by propriety?' " (Mahood, 1974, p. 305; see *AN* 17.9). A proper understanding of the rich meanings of *li* is a requisite for figuring out its role in Confucius' virtue education. I shall argue that *li* is best understood in two respects: outwardly, *li* refers to the various ritual norms prescribing what to do in different contexts; internally, *li* has to do with proper moral psychology such as feelings, motivations, and desires. In other words, *li* takes the form of various rituals and norms, whereas the substance of *li* is the proper spirit in which they are carried out.

For Confucius, if the rites are practised without going with the agent's proper feelings or attitudes, *li* is deadened or collapses. When Lin Fang asked about the "essence" of the rituals, the Master replied: "A great question indeed! In the rituals, frugality is preferred to extravagance; in mourning, deep sorrow is preferred to minute attention to observances" (*AN* 3.4). This is a key chapter to elaborating *li*. This grammar indicates that Confucius was trying to balance two values, that is, frugality and deep sorrow on the one hand and extravagance and observances on the other. When conducting the rituals, Confucius preferred frugality to extravagance. That is because he was afraid that people may misplace the value of practising *li* on the formality of the rituals. In that case, *li* becomes specious. However, the real value of the rituals does not lie in pompous observances but lies in genuine and sincere feelings (e.g., deep sorrow). The lack of sincerity of feelings and attitudes, or even the utter absence of proper feelings in performing a ritual, negates its point (Chong, 1998). A metaphor for this goes as follows: believing that, "Just like a lady must have natural good looks before cosmetics can enhance her beauty" (Huang, 1997, p. 60), Confucius said to Zi-xia: "Painting comes after a white ground". Zi-xia responded instantly: "So the rituals come after too?" (*AN* 3.8). The question is, after "what"? I contend that originally the rituals came into existence only after people had genuine feelings to express appropriately. This is supported by Mahood's idea:

> The Chinese recognise that there is in some sense a given datum in the form of a human nature with natural needs (for example, the natural impulse towards showing affection); but they also recognise that the universalisation of moral rules which exist to satisfy those need is a prerequisite. The merely personal "felt" experience requires the sanction of universalisability that *li* provides.
>
> (1974, pp. 306–307)

Mahood goes further to make a connection between the universalisation of human needs (such as showing affection) with the universalisability of *li*. The rituals can be seen as a natural outcome of the human need to express sincere feelings. When explaining that, "Painting comes after a white ground", Huang wrote that, "This in turn means figuratively that simplicity comes before refinement, or humanity comes before the rituals" (1997, p. 60). Simple and pure human feelings must come before the rituals are invented and practised. Afterwards, the practice of rituals is accompanied by the display of proper feelings or attitudes. It holds also in worship. "When offering sacrifices, he *felt* as if the spirits were present; when offering sacrifices to the gods, he *felt* as if the gods were present" (*AN* 3.12 [italics mine]). The Master's reverence/jing敬 is fully displayed in worship rituals. Likewise, when responding to Zi-you's query about filial piety, Confucius said, "Nowadays filial piety merely means being able to feed one's parents. Even dogs and horses are fed. Without reverence, how can you tell the difference?" (*AN* 2.7). Without sincere attitudes and genuine feelings, the conduct of *li* (e.g., worship rituals and codes of filial piety) is simply formalistic and empty.

Accordingly, Confucius thought highly of the sincerity of the ruler's feelings and attitudes. The Master said: "If a man is not *ren*/humane, what can he do with the rituals? If a man is not *ren*/humane, what can he do with music?" (*AN* 3.3). The man refers specifically to the sovereign (Huang, 1997) or an official in authority. If the leader is not originally courteous and deferential to his people, what can we expect him to do with the rituals as well as music at his disposal? The Master said, "Seated on high, he is not lenient; performing the rituals, he is not reverent; confronting mourning he is not grief-stricken—How can I bear to look at all this!" (*AN* 3.26). If the leader has no proper attitudes and feelings towards his people, "rites collapse and music is spoilt" imminently and the Master cannot bear to see it.

In sum, the Master said: " 'The rituals, the rituals,' they say. Do they merely refer to jade and silk?" (*AN* 17.9). Jade and silk were used as gifts in offering tribute or making friends; they were merely instrumental to showing reverence (Huang, 1997). The proper spirit of *li* is the kindly feelings and warm regards. For the young, the observance and practice of *li* in its various forms serves to shape their character by "motivating one to behave, desire, feel and act in the required ways" (Chong, 1998, p. 101).

Form and essence: *wén zhì bīn bīn*

As noted, *li* has two aspects: the form and the substance. Confucius stressed over and over again that the essence of *li* is the sincere attitudes and genuine feelings. With that said, it must be noted that Confucius deliberately avoided the risk of emphasising the essence/spirit/substance of *li* at the cost of the form of *li*. Rather, the Master said,

> When *zhi*/simplicity surpasses *wen*/refinement, one is a peasant; when *wen*/refinement surpasses *zhi*/simplicity, one is a pedant. Only when

> *wen*/refinement and *zhi*/simplicity are well blended (*wén zhì bīn bīn*) can one become a gentleman.
>
> (*AN* 6.18)

Zhi literally means unadorned nature, rustic substance; *wen* means refinement, adornment. A gentleman's *zhi* and *wen* are equally matched: in order to express his plain and simple feelings properly, the refined *li* comes to his aid or rather is at his disposal. This is the case because the gentleman's previous virtue education has equipped him with refined forms of *li* by means of which his earnest feelings can be properly expressed. A closely related chapter runs as follows:

> Ji Zi-cheng said: What the gentleman needs is simplicity, that is all. *What does he need refinement for?* Zi-gong said: What a pity that Your Excellency should have made such a remark about the gentleman! Even a team of four horses cannot overtake the tongue. *Refinement is as important as simplicity; simplicity is as important as refinement.* The hide of a tiger or a panther stripped of its hair is not any different from that of a dog or a sheep stripped of its hair.
>
> (*AN* 12.8 [italics mine])

Zi-gong's response reinforced his Master's point about the importance both of *wen* and *zhi*, the appropriate balance of *wen* with *zhi*, and the harmony of the refinement of *li* with the proper spirit of *li*.

The regulation of emotions by *li*

Once the spirit of *li* is abstracted and transformed into the various rituals, these rites and norms can act in turn to restrain one's feelings and prevent them from going to extremes. An interesting example shows up in *AN*, chapters 11.8–11.11 concerning Yan Yuan's death. When he died, the Master lamented him excessively saying: "Alas, Heaven is killing me! Heaven is killing me!" (*AN* 11.9). Confucius' followers said to him: "Sir, you are grieving to excess". The Master replied: "Am I? If I do not grieve for this man excessively, for whom else should I?" (*AN* 11.10). These passages demonstrate that Confucius believed Yan Yuan's death deserved his lament, and his deep sorrow for Yan Yuan was justified and appropriate.

Nevertheless, when Yan Yuan's father Yan Lu requested the Master's carriage to serve as Yan Yuan's outer coffin, Confucius declined. He gave two reasons (*AN* 11.8): first, when his own son, Bo-yu, died, he had only an inner coffin and no outer coffin; second, according to the rituals, a minister couldn't go on foot and must be carried by a carriage. Since Master Kong once served as a minister in Lu, he could not go on foot. Another event is this: when Yan Yuan's classmates wanted to bury him handsomely, the Master said, "That will not do" (*AN* 11.11). That is because as stipulated in the rituals, funeral expenses should accord with the financial status of the deceased's family (Huang, 1997). Yan Yuan came from a notoriously impoverished family, so the Master disapproved of his disciples' suggestion.

This case shows that even when Confucius was deeply grieved at his favourite disciple's death, he still stuck to the rituals and did not break *li* to bury Yan Yuan handsomely. If he had chosen to do so, he would have been criticised for letting emotions trample his decisions. That would be a defect in character: letting one's grief and sorrow go too far so as to miss *li*. In this case, Confucius' deep sorrow was expressed in a way in congruent with the restriction of *li*; specifically, funeral rituals. Confucius is a living example of his saying "To restrain oneself and return to the rituals constitutes *ren*/humanity" (*AN* 12.1). Moreover, this is an instance of "harmony" (in Huang's translation) or "natural ease" (in Legge's translation, 2015). As put by Master You, "In practicing the rituals, a natural ease is to be prized" (*AN* 1.12). That is, ideally, a natural ease or harmony is to exist between one's sincere feelings/*zhi* and the rituals/*wen*.

At this juncture, it needs to be noted that *li* is often translated as "ritual propriety". Propriety connotes the idea of what is proper (Legge, 2015). This translation implies that properness or righteousness is imbued or written into the rituals. The properness in question is *yi*. *Yi* "in its traditional interpretation, means the doing of what is yi [宜]-suitable, or appropriate" (Chong, 1998, p. 113). In short, *yi* is written into *li*; it explains why *li* is frequently translated as ritual propriety. Regarding the relation between *yi* and *li*, Confucius said: "The gentleman in everything considers righteousness (*yi*) to be essential. He performs it according to (*li*) the rules of propriety" (*AN* 15.18). *Yi* is integrated into *li*. For that matter, the ritual propriety itself is of great educational significance. The student who receives ritual education is starting to learn what is righteous or appropriate by first following the stipulated rituals. That is, in terms of the order of learning, the form of *li* comes first and the spirit of *li* and *yi* is normally to be appreciated later.

Moreover, it is plausible to conceive that *yi* is concerned with the appropriateness not only of outward behaviour, but also of inner feelings. That is, the two aspects of *li* – namely, *wen* and *zhi* – must be in congruent with *yi* and hence proper.

Ideal: harmony between the rituals and inner feelings

A well-known self-description of Confucius is as follows:

> At fifteen, I bent my mind on learning; at thirty, I was established; at forty, I was free from delusion; at fifty, I knew the decree of Heaven; at sixty, my ears became subtly perceptive; *at seventy, I was able to follow my heart's desire without overstepping the rules of propriety.*
>
> (*AN* 2.4 [italics mine])

Acting in accordance with one's inclinations without transgressing the rituals and norms is an ideal for Confucius. It is "the ideal of living naturally" (Sarkissian, 2010, p. 1) or rather, living at natural ease. It is worth noting that the inclinations in question are not inborn but formed by ritual education as discussed previously. Specifically, this natural ease refers to the harmony between Confucius' desires and the rules of propriety. At this final stage, Confucius' inner feelings and attitudes are in complete harmony with *li* and *yi*. Therefore, he can simply follow his heart's desire to hit the mark of *li* and *yi*. This is the case because Confucius' desires, feelings, and attitudes had been successfully disciplined and regulated by the ritual propriety, and consequently, his inclinations became appropriate, *yi*.

A case in point is as follows. Zai Wo asked Confucius about the three-year mourning ritual for parents, saying that one year was long enough (*AN* 17.19). The Master replied, "Eating rice and wearing brocade – do you feel at ease?" Zai Wo said: "I do". The Master said,

> If you do, then do so! When the gentleman is in mourning: eating delicacies, he does not relish their good taste; listening to music, he does not feel any happiness; living at home, he does not enjoy its comfort. Therefore, he does not do so. Now, if you feel at ease, do so!
>
> (*AN* 17.19)

At first, the Master suggested Zai Wo simply follow his heart. He wanted his disciple to see whether his innovative idea can pass the test of heart. And then Confucius elaborated on the stark difference between Zai Wo's current state of heart and that of a gentleman. Confucius intended to take a gentleman's heart as the benchmark against which Zai Wo should gauge the gap between himself and the gentleman and accordingly "correct" himself. Confucius conjectured that even if the ritual were de-regulated, the gentleman could not feel ease when enjoying delicious food, good music, or cosy shelter. As a result, he would rather retain the three-year mourning.

The cultivated inner feelings of an ideal character can act as a guide to his life. A useful passage is this: the Master said, "It is only the truly virtuous man, who can love, or who can hate, others" (*AN* 4.3). That is because his love and hatred is noble, appropriate and in accord with *yi*. Related to this, when Zi Gong asked, "What do you say of a man who is loved by all the people of his neighbourhood?", the master replied, "We may not for that accord our approval of him". "And what do you say of him who is hated by all the people of his neighbourhood?" The Master said, "We may not for that conclude that he is bad. It is better than either of these cases that the good in the neighbourhood love him, and the bad hate him" (*AN* 13.24). Clearly, Confucius regards the virtuous people's love and hatred as an objective standard of judgment. He would say a man who intends to please everyone, be he virtuous or not, in one's village, is a "thief of virtue"

(*AN* 17.11). He steals the name of virtue since what he really seeks is honour (false reputation) bestowed by villagers who are not qualified as good judges of virtue.

Implications for character education

The primacy of virtue and character is a definite mark of Confucius' ethics. Virtue and vice concepts are permanent in *AN*. Given that Confucius "makes virtues and vices a focal point of philosophical reflection about ethics" (Roberts, 2017, p. 18), his ethics could be conceived as a variety of virtue ethics. As stated previously, *AN* on the whole centres on the theme of self-cultivation, the prospective end result of which is a virtuous person, as well as a virtuous official. Therefore, character and virtue education naturally become primary concerns of Confucius' ethics. It should be clear from preceding passages that ritual education is the linchpin of Confucius' virtue curriculum. Even though Confucius' curriculum also included the study of music, poetry, archery, and history, all considered important to cultivating virtue (Sarkissian, 2010), Confucius' remarks on archery were intended to emphasise the link between archery and the rituals instead of the skills of archery itself (see *AN* 3.7 and 3.16). The core status that learning the rituals occupies in the curriculum is easy to see.

Li as a code of propriety has its outer forms and inner spirits. The rituals are expected to be performed in a proper spirit. As noted, the appropriate balance of form and essence (i.e., *wén zhì bīn bīn*) is the mark of a gentleman. With this ultimate aim in mind, Confucian character education starts with the observance of the various established rituals. These rituals are intended not only to establish one's norms of conduct in the various spheres of life, but also to elicit one's appropriate feelings or to restrain excessive natural affections when confronting associated situations. One's true feelings and affections need to be expressed, and it is paramount that they be expressed properly. The rituals provide people with pertinent occasions to learn to express emotions appropriately. That is, the rituals can be utilised for the educational purpose of disciplining one's feelings; specifically, evoking the proper affections and feelings (e.g., reverence to the ancestors, filial love for parents). This whole idea tallies with Plato's and Aristotle's thought as follows:

> For moral virtue is concerned with pleasures and pains. . . . Hence we ought to have been brought up in a particular way from our very youth, as Plato says, so as both to delight in and to be pained by the things that we ought; this is the right education.
>
> (Aristotle, 2009, p. 26; 1104b9–13)

For the young, the right education starts with having proper feelings with respect to appropriate objects. For that matter, the idea of *yi*/appropriateness/

rightness is also written into emotional education of this type. As previously discussed, the rituals first serve as the educational channel for eliciting pertinent emotions and then in turn act as an appropriate expression of the felt emotions. Throughout the process, moral exemplars, such as virtuous rulers and gentlemen, act as lively role models, demonstrating the match between outward behaviours and inner feelings.

This chapter can be concluded with a remark highlighting the great educational significance of the rituals for character education. The central role played by the rituals in Confucian virtue ethics and character education could be beneficially revived and used for the betterment of modern societies. Many rituals go along with festivals such as Teachers' Day (celebrated on Confucius' birthday), Mother's Day, Father's Day, and Tomb Sweeping Day. They are good occasions for virtue and character education centred on *li*, *yi*, and *ren*.

References

Aristotle (2004). *Rhetoric* (W. R. Roberts, Trans.). Mineola, NY: Dover Publications.

Aristotle (2009). *The Nicomachean ethics* (Rev. ed., D. Ross and L. Brown, Trans.). Oxford: Oxford University Press.

Chan, W.-T. (1955). The evolution of the Confucian concept Jên. *Philosophy East and West*, *4*(4), 295–319.

Chong, C. K. (1998). Confucius's virtue ethics. Li, Yi, Wen and Chih in the Analects. *Journal of Chinese Philosophy*, *25*, 101–130.

Cua, A. S. (1971). Reflections on the structure of Confucian ethics. *Philosophy East and West*, *21*(2), 125–140.

Flanagan, F. M. (2011). *Confucius, the Analects and western education*. London: Continuum.

Huang, C. (1997). *The Analects of Confucius (Lun Yu): A literal translation with an introduction and notes*. New York: Oxford University Press.

Kristjánsson, K. (2006). Emulation and the use of role models in moral education. *Journal of Moral Education*, *35*(1), 37–49.

Kristjánsson, K. (2007). *Aristotle, emotions, and education*. London: Routledge.

Legge, J. (2015). *The Analects of Confucius*. Taipei City, Taiwan: Business Weekly Publishing.

Luo, S. (2012). Confucius's virtue politics: Ren as leadership virtue. *Asian Philosophy*, *22*(1), 15–35.

Mahood, G. H. (1974). Human nature and the virtues in Confucius and Aristotle. *Journal of Chinese Philosophy*, *1*, 295–312.

Olberding, A. (2012). *Moral exemplars in the Analects: The good person is that*. New York: Routledge.

Plato (n.d.). *Protagoras* (B. Jowett, Trans.). Retrieved from http://classics.mit.edu/Plato/protagoras.html

Roberts, R. C. (2017). Varieties of virtue ethics. In D. Carr, J. Arthur, & K. Kristjánsson (Eds.), *Varieties of virtue ethics* (pp. 17–34). London: Palgrave Macmillan.

Sarkissian, H. (2010). Confucius and the effortless life of virtue. *History of Philosophy Quarterly*, *27*(1), 1–16.

Slingerland, E. (2001). Virtue ethics, the Analects, and the problem of commensurability. *Journal of Religious Ethics*, *29*(1), 97–125.

Tan, S.-H. (2005). Imaging Confucius: Paradigmatic characters and virtue ethics. *Journal of Chinese Philosophy*, *32*(3), 409–426.

Wong, W. (2011). The moral and non-moral virtues in Confucian ethics. *Asian Philosophy*, *21*(1), 71–82.

Yu, J. (1998). Virtue: Confucius and Aristotle. *Philosophy East and West*, *48*(2), 323–347.

Yu, J. (2007). *The ethics of Confucius and Aristotle: Mirrors of virtue*. London: Routledge.

Index

Page numbers in **bold** *indicate a table on the corresponding page.*